# Clear**Revise**

## AQA GCSE
## **Physics** 8463 / 8464

Illustrated revision and practice

Foundation and Higher
Physics and Trilogy Courses

Published by
PG Online Limited
The Old Coach House
35 Main Road
Tolpuddle
Dorset
DT2 7EW
United Kingdom

sales@pgonline.co.uk
www.clearrevise.com
www.pgonline.co.uk
**2021**

**PG** ONLINE

# PREFACE

Absolute clarity! That's the aim.

This is everything you need to ace the examined component in this course and beam with pride. Each topic is laid out in a beautifully illustrated format that is clear, approachable and as concise and simple as possible.

Each section of the specification is clearly indicated to help you cross-reference your revision. The checklist on the contents pages will help you keep track of what you have already worked through and what's left before the big day.

We have included worked exam-style questions with answers for almost every topic. This helps you understand where marks are coming from and to see the theory at work for yourself in an exam situation. There is also a set of exam-style questions at the end of each section for you to practise writing answers for. You can check your answers against those given at the end of the book.

# LEVELS OF LEARNING

Based on the degree to which you are able to truly understand a new topic, we recommend that you work in stages. Start by reading a short explanation of something, then try and recall what you've just read. This has limited effect if you stop there but it aids the next stage. Question everything. Write down your own summary and then complete and mark a related exam-style question. Cover up the answers if necessary but learn from them once you've seen them. Lastly, teach someone else. Explain the topic in a way that they can understand. Have a go at the different practice questions – they offer an insight into how and where marks are awarded.

# ACKNOWLEDGEMENTS

**The questions in the ClearRevise textbook are the sole responsibility of the authors and have neither been provided nor approved by the examination board.**

Every effort has been made to trace and acknowledge ownership of copyright. The publishers will be happy to make any future amendments with copyright owners that it has not been possible to contact. The publisher would like to thank the following companies and individuals who granted permission for the use of their images in this textbook.

Design and artwork: Jessica Webb & Mike Bloys / PG Online Ltd
Insulation image: Martyn F Chillmaid/Science Photo Library. Sciencephotos / Alamy Stock Photo.
Resistance experiment image: © iStock. Frictionless ramp image: © Adobe Stock. Icons: Icons8.com
All other images: © Shutterstock
Contributor: Helen Sayers
Editing: Jim Newall

First edition 2021. 10 9 8 7 6 5 4 3 2 1
A catalogue entry for this book is available from the British Library
ISBN: 978-1-910523-33-9
Copyright © PG Online 2021
All rights reserved

Printed on FSC certified paper by Bell and Bain Ltd, Glasgow, UK.

# THE SCIENCE OF REVISION

## Illustrations and words

Research has shown that revising with words and pictures doubles the quality of responses by students.[1] This is known as 'dual-coding' because it provides two ways of fetching the information from our brain. The improvement in responses is particularly apparent in students when asked to apply their knowledge to different problems. Recall, application and judgement are all specifically and carefully assessed in public examination questions.

## Retrieval of information

Retrieval practice encourages students to come up with answers to questions.[2] The closer the question is to one you might see in a real examination, the better. Also, the closer the environment in which a student revises is to the 'examination environment', the better. Students who had a test 2–7 days away did 30% better using retrieval practice than students who simply read, or repeatedly reread material. Students who were expected to teach the content to someone else after their revision period did better still.[3] What was found to be most interesting in other studies is that students using retrieval methods and testing for revision were also more resilient to the introduction of stress.[4]

## Ebbinghaus' forgetting curve and spaced learning

Ebbinghaus' 140-year-old study examined the rate in which we forget things over time. The findings still hold power. However, the act of forgetting things and relearning them is what cements things into the brain.[5] Spacing out revision is more effective than cramming – we know that, but students should also know that the space between revisiting material should vary depending on how far away the examination is. A cyclical approach is required. An examination 12 months away necessitates revisiting covered material about once a month. A test in 30 days should have topics revisited every 3 days – intervals of roughly a tenth of the time available.[6]

## Summary

Students: the more tests and past questions you do, in an environment as close to examination conditions as possible, the better you are likely to perform on the day. If you prefer to listen to music while you revise, tunes without lyrics will be far less detrimental to your memory and retention. Silence is most effective.[5] If you choose to study with friends, choose carefully – effort is contagious.[7]

1.  Mayer, R. E., & Anderson, R. B. (1991). Animations need narrations: An experimental test of dual-coding hypothesis. *Journal of Education Psychology*, (83)4, 484–490.

2.  Roediger III, H. L., & Karpicke, J.D. (2006). Test-enhanced learning: Taking memory tests improves long-term retention. *Psychological Science*, 17(3), 249–255.

3.  Nestojko, J., Bui, D., Kornell, N. & Bjork, E. (2014). Expecting to teach enhances learning and organisation of knowledge in free recall of text passages. *Memory and Cognition*, 42(7), 1038–1048.

4.  Smith, A. M., Floerke, V. A., & Thomas, A. K. (2016) Retrieval practice protects memory against acute stress. *Science*, 354(6315), 1046–1048.

5.  Perham, N., & Currie, H. (2014). Does listening to preferred music improve comprehension performance? *Applied Cognitive Psychology*, 28(2), 279–284.

6.  Cepeda, N. J., Vul, E., Rohrer, D., Wixted, J. T. & Pashler, H. (2008). Spacing effects in learning a temporal ridgeline of optimal retention. *Psychological Science*, 19(11), 1095–1102.

7.  Busch, B. & Watson, E. (2019), *The Science of Learning*, 1st ed. Routledge.

# CONTENTS

## Topic 3     Particle model of matter

## Topic 4     Atomic structure

## Topic 8    Space physics

**Physics** ■    **Trilogy** ■      ☑

# MARK ALLOCATIONS

**Green mark allocations[1]** on answers to in-text questions throughout this guide help to indicate where marks are gained within the answers. A bracketed '1' e.g.[1] = one valid point worthy of a mark. In longer answer questions, a mark is given based on the whole response. In these answers, a tick mark[✓] indicates that a valid point has been made. There are often many more points to make than there are marks available so you have more opportunity to max out your answers than you may think.

# TOPICS FOR PAPER 1
Energy, electricity, particle model of matter and atomic structure

## Information about Paper 1:

### Separate Physics 8461:

**Written exam: 1 hour 45 minutes**
**Foundation and Higher Tier**
**100 marks**
**50% of the qualification grade**
**All questions are mandatory**

### Trilogy 8464:

**Written exam: 1 hour 15 minutes**
**Foundation and Higher Tier**
**70 marks**
**16.7% of the qualification grade**
**All questions are mandatory**

#### Specification coverage

The content for this assessment will be drawn from Topics 1–4  Energy; Electricity; Particle model of matter; and Atomic structure.

#### Questions

A mix of calculations, multiple-choice, closed short answer and open response questions assessing knowledge, understanding and skills.

Questions assess skills, knowledge and understanding of Physics.

# ENERGY STORES AND SYSTEMS

**Energy** is stored in **systems**. A system is an object or a group of objects.

## Common energy stores

Common energy stores are: **chemical**, **kinetic**, **elastic potential**, **gravitational potential** and **thermal**. Further stores are **magnetic**, **electrostatic** and **nuclear**. When a system changes, the way the energy is stored in the system changes.

| Scenario | Store that decreases | Store that increases |
|---|---|---|
| An object thrown upwards | Kinetic | Gravitational potential |
| A moving object hitting a wall | Kinetic | Thermal |
| A bow releasing an arrow | Elastic potential | Kinetic |
| A vehicle slowing down when braking | Kinetic | Thermal (caused by friction) |
| Heating water on a camping stove | Chemical (in gas) | Thermal |
| Using a battery-operated fan | Chemical (in battery) | Kinetic and thermal (of the motor and surroundings) |

## Change in systems

Systems can be changed by:

### Heating

### Work done by forces

### Work done when current flows

1. Complete the sentences using answers from below.

   **chemical    elastic potential    gravitational potential
   kinetic    thermal**

   (a) A cyclist accelerates. Energy is transferred from a
   chemical store to a _____ store. [1]

   (b) When a stone falls from a cliff, energy is transferred
   from a _____ store to a
   _____ store. [2]

2. A fully charged electric scooter is ridden down a street
   at a constant speed. Explain how the energy stores
   change in this system. [3]

*1. (a) kinetic[1], (b) gravitational potential[1], kinetic[1]*

*2. Three points from: Chemical energy[1] stored in the
   battery is transferred to the thermal energy[1] store of the
   scooter and then to the thermal energy[1] store of the
   surroundings due to air resistance and friction between
   the wheels and the ground[1].*

# KINETIC ENERGY ($E_k$)

## Calculating kinetic energy

A moving object has a store of **kinetic energy**. The kinetic energy of a moving object can be calculated using the equation:

Kinetic energy = 0.5 × mass × (speed)², or $E_k = \frac{1}{2}mv^2$

You need to be able to recall and apply this equation.

$E_k$ = kinetic energy in joules, J     $m$ = mass in kg   $v$ = velocity (speed) in metres per second.

The equation can be rearranged to calculate mass and speed.

1. A cyclist and bike have a combined mass of 75 kg. The cyclist is moving at 10 m/s. Calculate the kinetic energy of the cyclist and bike. [2]

2. A ball has a mass of 400 g. The ball moves with a speed of 2 m/s.

   Calculate the kinetic energy of the ball. Give the unit. [4]

3. **Higher only:** The mass of a runner is 70 kg. The kinetic energy of the runner is 875 J.

   Calculate the speed of the runner in m/s. [3]

Remember to convert any values that have been given in different units.

**For example:**

1 MJ = 1 000 000 J

0.1 kg = 100 g

1 km = 1000 m; 1 cm = 0.01 m

1 minute = 60 seconds;
1 hour = 60 × 60 = 3600 seconds.

1. $E_k = \frac{1}{2}mv^2$
   = 0.5 × 75 × 10²[1]
   = 3750 J[1]

2. $m$ = 400 g = 0.4 kg[1]
   $E_k = \frac{1}{2}mv^2$
   = 0.5 × 0.4 × 2²[1]
   = 0.8[1] J[1]

3. $E_k = \frac{1}{2}mv^2$
   875 = 0.5 × 70 × v²[1]
   $v = \sqrt{\frac{2 \times 875}{70}}$[1]
   $v$ = 5 m/s[1]

# ELASTIC POTENTIAL ENERGY ($E_e$)

A stretched object, such as a spring or an elastic band under tension, stores **elastic potential energy**.

## Calculating elastic potential energy

The elastic potential energy of an object can be calculated using the equation:

elastic potential energy = 0.5 × spring constant × (extension)²

$$E_e = \frac{1}{2} ke^2$$

You need to be able to select this equation from the equation sheet and apply it.

$E_e$ = elastic energy in joules, J

$k$ = spring constant in newtons per metre, N/m

$e$ = extension in metres, m

The equation can be rearranged to calculate the spring constant and extension.

A spring also has a store of elastic potential energy if it is squashed. The amount of energy can be calculated using the same equation, but with compression replacing extension.

## The limit of proportionality

The **limit of proportionality** refers to the point at which an object will no longer return to its original shape or length when stretched or squashed. Beyond this point, you cannot calculate the elastic potential energy. See **page 72**.

1. A toy pops out on a compressed spring when a box is opened.
   (a) Describe the change in energy stores as the box opens. [2]
   (b) The spring extends by 0.1 m when the box opens. The spring constant is 4.0 N/m .
       Calculate the elastic potential energy, $E_e$, (J) stored in the spring when it is inside the box. [2]
2. **Higher only:** A training band stretched by 0.05 m has an elastic potential energy store of 0.02 J.
   Calculate the spring constant of the band in N/m. [3]

1. (a) *Elastic potential energy*[1] *is transferred to kinetic energy*[1].
   (b) $E_e = \frac{1}{2} ke^2$
   $\quad\quad = 0.5 \times 4.0 \times 0.1^{2[1]} = 0.02 \text{ J}^{[1]}$

2. $E_e \quad = \frac{1}{2} ke^2$
   $\quad 0.02 = 0.5 \times k \times 0.05^{2[1]}$
   $\quad\quad k \quad = \dfrac{2 \times 0.02}{0.05^2}$ [1]
   $\quad\quad\quad = 16 \text{ N/m}^{[1]}$

# GRAVITATIONAL POTENTIAL ENERGY ($E_p$)

When an object is raised it gains **gravitational potential energy (g.p.e.)**.

## Calculating gravitational potential energy

The gravitational potential energy gained by an object raised above the ground can be calculated using the equation:

g.p.e. = mass × gravitational field strength × height

$$E_p = mgh$$

You need to be able to recall and apply this equation.

$E_p$ = gravitational potential energy in joules, J

$m$ = mass in kilograms, kg

$g$ = gravitational field strength in newtons per kilogram, N/kg

$h$ = height in metres, m.

In any calculation the value of the gravitational field strength ($g$) will be given. Remember that $h$ is the change in **vertical** height.

The equation can be rearranged to calculate mass, gravitational field strength and height.

Use $g$ = 9.8 N/kg for all questions.

1. Calculate the increase in gravitational potential energy when a mass of 1.2 kg is lifted to a height of 5 m. [2]

2. A skier has a mass of 80 kg. Calculate how much energy is transferred from the gravitational potential energy store as the skier travels 150 m down a mountain. [2]

3. A ball is dropped from a height of 2.0 m. Calculate the speed at which the ball hits the ground. [3]

1. g.p.e. = $mgh$
   = 1.2 × 9.8 × 5[1]
   = 58.8 J[1]

2. g.p.e. = $mgh$
   = 80 × 9.8 × 150[1]
   = 117 600 J[1]

3. decrease in g.p.e. = increase in kinetic energy
   $m × g × h$ = 0.5 × $m$ × $v^2$[1]
   0.5 × $v^2$ = 9.8 × 2.0[1]
   $v = \sqrt{\dfrac{19.6}{0.5}}$ = 6.3 m/s[1]

# ENERGY CHANGES IN SYSTEMS

## Specific heat capacity

The amount of **thermal energy** needed to increase the temperature of a substance depends on its mass and **specific heat capacity**.

The specific heat capacity of a substance is the amount of energy needed to increase the temperature of one kilogram (1 kg) of the substance by one degree Celsius (1 °C).

When a substance is heated, there is an increase in its thermal energy store. When a substance cools down, there is a decrease in its thermal energy store. The change in the thermal energy store when the temperature is increased is the same as the change in the thermal energy store when the temperature decreases by the same amount.

The change in thermal energy store as the temperature of a system changes, can be calculated using the equation:

change in thermal energy = mass × specific heat capacity × temperature change

$$\Delta E = m\,c\,\Delta\theta$$

> $\Delta$ means there is a change. You need to be able to select this equation from the equation sheet and apply it.

$\Delta E$ = change in thermal energy in joules, J

$m$ = mass in kilograms, kg

$c$ = specific heat capacity in joules per kilogram per degree Celsius, J/kg °C

$\Delta\theta$ = temperature change in degrees Celsius, °C

The equation can be rearranged to calculate mass, specific heat capacity, and temperature change.

1. The specific heat capacity of copper is 385 J/kg °C

   Calculate the change in thermal energy when a 2.5 kg block of copper cools down by 20 °C.

   Give your answer in kJ.                                                              [3]

2. In an investigation to find the specific heat capacity of water, the temperature of 0.2 kg of water increased by 30 °C. The energy transferred to the thermal energy store of the water was 25.8 kJ.

   Calculate the specific heat capacity of water.                                       [3]

1. $\Delta E = mc\Delta\theta$

   $= 2.5 \times 385 \times 20$ [1]

   $= 19\,250$ J [1]

   $= 19.25$ kJ [1]

2. $\Delta E = mc\Delta\theta$

   $25\,800 = 0.2 \times c \times 30$ [1]

   $c = \dfrac{25\,800}{0.2 \times 30}$ [1]

   $= 4300$ J/kg °C [1]

# POWER

## Calculating power

**Power** is the **rate** at which energy is transferred, or the rate at which work is done. This means power is the amount of energy transferred, or how much work is done, in any given amount of time.

$$power = \frac{Energy\ transferred}{time} = \frac{Work\ done}{time}$$

$$P = \frac{E}{t} = \frac{W}{t}$$

$P$ = power in watts, W

$E$ = energy transferred in joules, J

$W$ = work done in joules, J

$t$ = time in seconds, s

**1 watt = 1 joule per second**

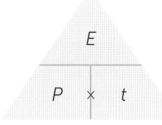

You could use $W$ instead of $E$ in the formula triangle.

Work done is represented by an italicised $W$ so be careful not to muddle it with the unit watts, W.

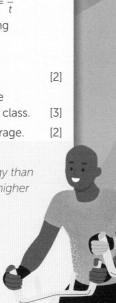

1. Choose the equation used to calculate work done. Tick **one** box. [1]

   □ $W = Pt$        □ $W = \frac{P}{t}$        □ $W = \frac{E}{t}$

2. Sara and Rob both do the same 45-minute cycling spin class.

   (a) Sara does more work than Rob in the class. Describe what this means in terms of power. [2]

   (b) Sara does 594 000 J of work in the 45-minute class. Calculate her average power during the class. [3]

   (c) Explain why the power is calculated as an average. [2]

   *1.  W = Pt[1]*

   *2. (a) Sara does more work/transfers more energy than Rob in the same time[1] so Sara's power is higher than Rob's power in this class[1].*

   *(b) t  = 45 × 60 = 2700s[1]*

   *P = $\frac{W}{t}$*

   *P = 594 000 / 2700[1]*

   *= 220 Watts (W)[1]*

   *(c) The amount of energy transferred by Sara each second will not be constant[1] so the power/rate of transfer also changes and an average is calculated for the time given[1].*

# REQUIRED PRACTICAL 1 (14)

## Specific heat capacity

This activity helps you to work safely and accurately in calculating the specific heat capacity of a material or substance.

The change in the **thermal energy store** of a substance ($\Delta E$) must be linked with a measurable change in another energy store.

For greater accuracy it is important to minimise unwanted energy transfers. It is assumed that all of the energy is transferred to the substance being tested.

This practical involves heating up a substance with an electric heater. The **mass** ($m$) and the **temperature change** ($\Delta\theta$) of the substance are measured and recorded. The amount of energy supplied (the electrical **work done**) by the heater ($\Delta E$) can be calculated. This could also be measured using a device called a joulemeter.

The **specific heat capacity** ($c$) can then be calculated using: $c = \dfrac{\Delta E}{m\,\Delta\theta}$

### Finding the specific heat capacity of a solid metal

1. Measure the mass of the metal block in kilograms.

2. Heat the block using an electric heater placed inside a hole in the block.

3. Record the temperature of the block every minute for 10 minutes.

4. Calculate the work done by the heater using $\Delta E = P \times t$ (See **pages 7 and 30**) where $P$ is the power of the heater in watts, and $t$ is the time in seconds the heater has been on.

5. Plot a graph of $\Delta\theta$ against $\Delta E$ and draw a line of best fit.

6. The gradient of the straight part of the line $= \Delta\theta/(\Delta E)$.

7. $c = \dfrac{1}{\text{gradient} \times m}$

**Note:** The calculated specific heat capacity is often higher than the actual value. This is because not all of the energy supplied by the heater will be transferred to the substance. Some will be transferred to the thermal energy store of the heater and thermometer and some will be transferred to the thermal energy store of the surroundings. See **pages 10–11**.

### To reduce errors

- Insulate the metal thoroughly to reduce thermal energy transfer away from the metal.
- Add oil or water into the thermometer hole for better thermal contact.
- Use a datalogger to record the temperature and time simultaneously. This can also help to reduce errors reading the thermometer scale.
- Use a variable resistor to keep the heater current constant.

### Safety factors

- Place the metal block on a heat proof mat.
- Leave the heater to cool before handling.

Avoid using the word 'amount'.

Be specific and use the terms 'volume' or 'mass'.

If the power of the heater is not known, $\Delta E$ can be found by recording the potential difference, $V$, across the heater in volts, and the current, $I$, through it, in amps.

So, $\Delta E = V \times I \times t$. See **page 31**.

1. Name the equipment used to measure:

   (a) Mass. [1]    (b) Temperature. [1]

2. Give **two** safety factors to consider when investigating the specific heat capacity of water. [2]

3. Describe an experiment to find the specific heat capacity of olive oil in J/kg °C using an immersion heater and a joulemeter.

   Include the other apparatus used, ways to reduce errors and how to calculate the specific heat capacity of olive oil. [6]

1. *(a) Balance.[1]    (b) Thermometer.[1]*

2. *Any two from: care handling hot heater / hot water[1]; keep water away from electrical supply[1]; place the metal block on a heat proof mat[1]; leave the heater to cool before handling[1].*

3. *Place a beaker on a balance and zero it.[√] Add some oil to the beaker[√] record the mass of the oil in kg[√]. Remove the beaker from the balance.[√] Connect the joulemeter and immersion heater in a series circuit with a power supply and a switch.[√] Put the immersion heater into the oil[√] ensure it is fully submerged, so all the thermal energy is transferred to the oil[√]. Add a thermometer to the oil.[√] Insulate the beaker[√] and add a lid to reduce heat loss to the surroundings[√]. Record the starting temperature of the oil[√] and check that the joulemeter is reading zero. Switch on the circuit to begin heating the oil.[√] Stir the oil to ensure the thermal energy is evenly distributed.[√] Record the temperature and reading on the joulemeter after about 20 minutes.[√]*

   *Calculate the specific heat capacity using:*

   $$c = \frac{\Delta E}{m\,\Delta\theta}\ [√]$$

   *$\Delta E$ is the reading from the joulemeter in J.[√]*

   *$\Delta\theta$ = final temperature - starting temperature in °C.[√]*

   *m is the mass of oil heated in kg.[√]*

   *This is an extended response question that should be marked in accordance with the levels based mark scheme on page 170.*

# ENERGY TRANSFERS IN A SYSTEM

**Energy can only be stored, transferred or dissipated (spread out to the surroundings). It cannot be created or destroyed.**

Energy is transferred from one store into one or more other stores. Sometimes the energy is transferred in a **useful** way. When energy is described as **wasted** it means it has been transferred less usefully. It has been **dissipated**.

## Systems

Energy within a **closed system** transfers between stores keeping the net energy change in the closed system at zero. This means that no energy is dissipated or wasted from the system and the total amount of energy in the system is constant.

### Example system:

A hot drink is in an insulated cup with a lid. Energy is transferred from the thermal energy store of the drink to the cup and lid. Energy is then dissipated to the thermal energy store of the surroundings. The energy is less useful and cannot be gathered back together again.

There is no net change in energy in the system.

Usually we have to include the surroundings in a system to consider it as a closed system. Very few systems are completely isolated from their surroundings.

1. A playground swing is released with a child sitting still on it. It swings a few times and eventually stops.

    Describe the motion of the swing in terms of energy transfers. [5]

    *Before the swing is released it has a store of gravitational potential energy.[1] As the swing moves, energy is transferred to the kinetic energy store which reaches a maximum at the bottom of the swing.[1] As the swing moves upwards, energy is transferred to the gravitational potential energy store, and the kinetic energy store of the swing decreases.[1] The swing goes less high each time until it stops[1] because energy is dissipated to the surroundings due to the friction of the swing/child (with the air and where the swing is attached)[1].*

## Reducing unwanted energy transfers

Unwanted energy transfers happen for many reasons. It is useful to look for ways to reduce these, and therefore transfer energy as usefully as possible.

Any moving parts in a machine will waste energy because of resistance, or friction. **Friction** can be **reduced** by lubricating these moving parts using oils and powders for example.

When an electric current flows through a resistor, energy in the chemical store of a battery is transferred to the thermal energy store of the surroundings by electrical work.

When a device transfers energy to a thermal store, such as water in a kettle, the energy has been transferred to a useful store. However, energy is often wasted by devices dissipating energy to the thermal store of the surroundings. Good **thermal insulation** can reduce this loss. For example, insulation around a hot water tank reduces the amount of energy transferred to the thermal energy store of the surroundings.

Materials with a higher **thermal conductivity** have a higher rate of energy transfer by **conduction** across the material.

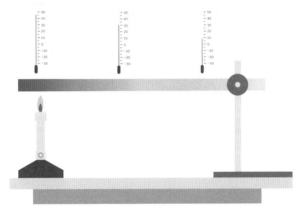

Good **conductors**, such as metals, transfer energy easily by the movement of free electrons. Materials with a low thermal conductivity are good **insulators**.

To prevent heat loss from buildings, the rate of cooling through the walls can be reduced by using insulating materials with thicker layers.

2. Complete the sentence:

   Thermal energy travels from a _____ place to a _____ place. [1]

3. New houses have walls with a low thermal conductivity. Explain how this helps to keep the inside of the building warm in winter. [2]

4. Suggest why the wires used to connect an electrical device to the mains supply should have a very low electrical resistance. [2]

   *2. warm; cooler.[1]*

   *3. A low thermal conductivity means that the rate of energy transfer is low.[1] This means that the rate of energy transfer from the inside to the outside of the building is reduced[1] so not as much energy is needed to keep the inside warm.*

   *4. A lower resistance means that less energy is transferred by electrical work,[1] so less energy is transferred to the thermal energy store of the surroundings[1].*

# REQUIRED PRACTICAL 2

## Thermal insulation

This practical activity helps you draw conclusions about the effectiveness and properties of different materials as thermal insulators.

### The effectiveness of different materials as thermal insulators

Hot liquids cool down as energy is transferred to the thermal energy store of the surroundings.

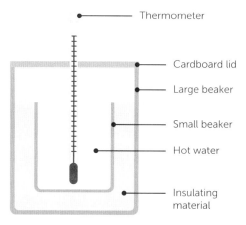

- Thermometer
- Cardboard lid
- Large beaker
- Small beaker
- Hot water
- Insulating material

The rate that energy is transferred from the water for each insulator can be compared by plotting **cooling curves**. These are graphs of temperature against time.

**Suggested method:**

1. Put hot water from a kettle into a 100 ml beaker.
2. Place the small 100 ml beaker inside a large beaker.
3. Place a cardboard lid on the large beaker with a hole for a thermometer.
4. Insert the thermometer through the lid into the hot water.
5. Record the temperature of the water and start the stopwatch.
6. Record the temperature of the water every 3 minutes for 15 minutes.
7. Repeat steps 1 to 6, filling the space between beakers with different insulating materials.
8. Draw cooling curves for each insulator.

Use the graphs to determine which material is the best insulator.

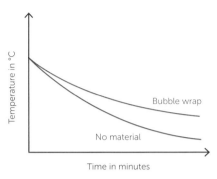

For the same amount of time, the material with the smallest temperature drop transfers energy the most slowly, so it is the best insulator.

The type of insulator used will affect how quickly energy is transferred from the beaker of water. Using insulators with air pockets reduces energy transfers because air has a very low thermal conductivity.

There are large gaps between the particles in trapped air which make it difficult for energy to be transferred.

**Always look for ways to reduce errors to make the results more valid.**

It is difficult to get the same starting temperature of hot water for each material but errors can be caused if it is different. Other errors could be introduced by:

- using the beakers at different temperatures
- putting the beakers on different surfaces/ next to a window
- incorrectly reading the scale on the thermometer
- stirring the water for different amounts of time.

## Investigating how the thickness of a material affects thermal insulation

A similar method can be used to the one for testing different materials.

Instead of using two beakers, wrap increasing numbers of layers of the same material around the beaker with hot water, holding it in place with an elastic band.

Adding layers of an insulator traps more air. This makes it harder for energy to be transferred to the thermal energy store of the surroundings and improves the insulation.

---

An experiment is done to find the effectiveness of different materials as thermal insulators.

(a) Name the independent variable. [1]     (b) Name the dependent variable. [1]

(c) Ideally the starting temperature of the water should be the same each time.
Give **two** other control variables in this investigation. [2]

(d) (i)  Give the resolution of this thermometer. [1]

-10   0   10   20   30   40   50   60   °C

(ii) Give **one** reason why using a temperature probe could improve the accuracy of the results. [1]

*(a) Time.[1]     (b) Temperature.[1]*

*(c) Volume of water[1], thickness / mass / number of layers of each material[1].*

*(d) (i) 1°C[1]*

*(ii) Any **one** from: temperature probe avoids human error in reading thermometer/probe does not need to be removed from the water to read value.[1]*

# EFFICIENCY

**Efficiency** shows how much of the total energy transferred is transferred to useful energy stores, and how much is transferred to less useful stores, or wasted.

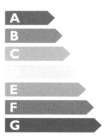

## Calculating efficiency

More efficient appliances, homes and vehicles cost less to run as more of the energy is transferred usefully. This can reduce carbon dioxide emissions if fossil fuels are used.

Efficiency can be calculated using the equations:

$$\text{efficiency} = \frac{\text{useful output energy transfer}}{\text{total input energy transfer}} = \frac{\text{useful power output}}{\text{total power input}}$$

Efficiency is given as a decimal value between 0 and 1.0, or as a percentage % (decimal × 100). Efficiency does not have a unit.

Power is the rate at which energy is transferred in watts (W) so the alternative equation above can be used.

You need to be able to recall and apply these equations.

### Higher Tier only

The efficiency of an energy transfer is increased when the ratio of useful energy to wasted energy is increased. The amount of useful energy output transfer is increased, and the amount of energy wasted is reduced.

For example, a light bulb can be made more efficient by reducing the amount of energy transferred by heating. Reducing friction by lubricating moving parts in a machine also increases the efficiency.

1. A lamp is rated at 12 W and transfers 4.8 W as visible light.
   Calculate the efficiency of the light bulb. [2]
2. A solar panel is 23% efficient. Calculate the energy wasted when 500 J is input. [3]
3. Explain why it is not possible to have an efficiency greater than 1.0. [2]
4. **Higher Tier only:** Double glazed windows are made of two panes of glass with a layer of gas between them. Explain why double glazing is more thermally efficient than a single pane of glass. [2]

*1. Efficiency = $\frac{\text{useful power output}}{\text{total power input}}$*

*= $\frac{4.8[1]}{12}$*

*= 0.4 or 40%[1]*

*2. Efficiency = $\frac{\text{useful output energy transfer}}{\text{total input energy transfer}}$*

*0.23 = $\frac{\text{useful output energy transfer}[1]}{500}$*

*useful output energy transfer = 0.23 × 500*

*= 115 J[1]*

*energy wasted = 500 − 115 = 385 J[1]*

*3. This would mean that the amount of useful energy output is greater than the total input energy transferred[1] which is not possible as energy cannot be created[1].*

*4. The gas is a poor thermal conductor and is trapped between two layers of glass so the thermal conductivity of the window is lower.[1] The rate of energy transfer is lower than for the single pane, so it is more thermally efficient.[1]*

# NATIONAL AND GLOBAL ENERGY RESOURCES

**Energy resources** store a large amount of energy. They can be used to transfer energy from one store to another, usually by generating electricity, and also for heating and transport.

| Non-renewable energy | Renewable energy |
|---|---|
| A non-renewable energy resource is a resource that is being used up faster than it can form. It has a finite supply. This is either because it forms very slowly or it is no longer being formed. It will eventually run out when all the reserves have been used up. | A renewable energy resource is one that can be replenished as it is used, either by human action or natural processes. |

### Example

Burning fossil fuels releases carbon dioxide into the atmosphere which contributes to global warming and climate change. However, some countries are heavily dependent on the distribution of fossil fuels, so they resist change to alternative resources.

Plans for tidal barrages have been rejected for the Severn Estuary because of potential environmental damage to marine habitats, but mainly because of the huge costs. Offshore floating wind farms are now proposed as an alternative to the tidal barrage to increase the output from renewables.

Nuclear fuel does not produce carbon dioxide but does create radioactive waste which raises issues of disposal and the safety of nuclear power stations.

1. Circle the non-renewable energy resources.                                                                    [1]

   Biofuel     Coal     Gas     Geothermal     Wind     Oil     Sun     Nuclear fuel

2. Hydroelectric, tidal and wave generators are three renewable energy resources that use water to produce electricity. Complete the table below showing information about the resources.     [3]

| | Energy store | Power output | Reliability | Environmental impact |
|---|---|---|---|---|
| Hydroelectric power | Gravitational potential energy | High | Reliable | |
| Tidal power | Gravitational potential energy | High | | Damage to marine and estuary habitats |
| Wave power | | Low | Reliable | Has a low impact |

*1. Coal   Gas   Nuclear Fuel   Oil[1].*

*2. Hydroelectric power – Flooding (dams) damages land / habitats[1]; Tidal – Reliable[1]; Wave – Kinetic.[1]*

# IMPACTS AND USES OF ENERGY RESOURCES

## Non-renewable energy sources

| Energy resource | Advantages | Disadvantages |
|---|---|---|
| Fossil fuels<br><br>Oil    Gas    Coal | Relatively low cost, easy to use for transport, very adaptable | Produce carbon dioxide (contributes to global warming) and sulfur dioxide (causes acid rain and can cause breathing problems; resources are finite (will run out). |
| Nuclear<br><br>Nuclear | No greenhouse gases produced. A large amount of energy produced from a small amount of fuel | Disposal of uranium is costly and hazardous. Huge cost to build, run and decommission power stations. Risk of catastrophic accidents. |

## Renewable energy sources

| Energy resource | Advantages | Disadvantages |
|---|---|---|
| Biofuel | Can use up plant and animal waste. Replacement engine fuel for transport. | Some biofuels are grown especially which uses a lot of land that could be used for food production. |
| Geothermal | Can be used for heating local areas. Electricity can be generated from hot rocks. | Not available everywhere and sometimes in dangerous or inaccessible areas. |
| Hydroelectricity | Available at any time and can be started and stopped quickly. | Areas of natural beauty can be flooded to make the dams and reservoirs required. |
| Sun | Solar cells on roof tops and solar farms generate electricity. Also used for heating water. Solar power stations are in development. | Not available at night. Works better in places with more intense sunlight. Visual impact on the landscape raises objections. |
| Tides | Electricity can be generated at predictable times. | Could damage marine and estuary habitats. Tends to be extremely expensive with few suitable places for barrages. |
| Water waves | Can generate electricity on a small scale. | Maintenance, impact on marine environment and inconvenience to shipping. |
| Wind | Many suitable offshore sites in the UK. Fairly low maintenance when established. Turbines can be built out at sea. | Unreliable unless the wind speed is within the right range. Visual pollution of turbines on land. Many turbines are needed to generate suitable amounts of electricity. |

# EXAMINATION PRACTICE

01  Choose the correct equation to calculate kinetic energy. Tick (✓) **one** box.                                                    [1]

☐  $E_k = \frac{1}{2} mv^2$

☐  $E_k = ke^2$

☐  $E_k = \frac{1}{2} ke$

02  The diagram shows a spring before and after it is stretched.

Calculate the value of $e$, the spring extension, when it is stretched.                                                              [1]

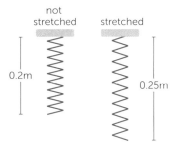

03  Calculate the kinetic energy store of a bird with a mass of 0.4 kg flying at 8 m/s. Give the unit.    [3]

04  A mountain biker rides along a path and jumps into the air. The cyclist and bike have a total mass of 85 kg.

04.1  Calculate the gravitational potential energy when the cyclist has jumped 2 m above the ground. Use $g$ = 9.8 N/kg.    [2]

04.2  Describe the energy transfers as the bike falls and then hits the ground.    [2]

04.3  **Higher Tier only:** The cyclist does another jump and has a maximum gravitational potential energy of 680 J. Calculate the speed at which the cyclist hits the ground.

Assume that there are no energy transfers due to friction or other forces.    [3]

05  The table below shows diesel and electric car sales in the UK for the same month in 2018 and 2019.

| Year | Diesel car sales | Electric car sales |
|------|------------------|--------------------|
| 2018 | 51 108 | 9971 |
| 2019 | 36 941 | 15 132 |

Identify the trends in both car sales. Suggest a reason for each trend.                    [4]

Diesel car sales:                              Electric car sales:

Trend:                                         Trend:

Reason:                                        Reason:

06  Give **one** advantage and **one** disadvantage of using nuclear fuel instead of coal to generate electricity.                                                                            [2]

07  The thermal conductivity of brick is approximately six times higher than the thermal conductivity of air.

Explain why the walls of houses are often built using two layers of bricks with a gap in between them, rather than a single layer of bricks.                                              [3]

08  A device with a total input power of 30 W has an efficiency of 0.45.

Calculate how much **useful energy** the device transfers in 20 s                           [3]

09  Explain why a more powerful heater is better at warming up a cold room than a less powerful heater.                                                                                      [2]

10  The specific heat capacity of water is 4200 J/kg°C

10.1  Define the specific heat capacity of water.                                           [1]

A student does an experiment to find the specific heat capacity of water using an electric heater with a power of 30 W.

10.2  What does the student need to measure to find the energy transferred by the heater?   [1]

The student completes the experiment and finds the specific heat capacity of water to be 4500 J/kg°C. All measurements and calculations were done accurately and repeated.

10.3  Suggest how the experiment could be improved to get a value nearer to 4200 J/kg°C.     [2]

10.4  **Higher Tier only:** Calculate the expected rise in temperature when 2100 J is transferred to 250 g of water. Use $c$ = 4200 J/kg°C.                                               [4]

11  **Physics only:** Describe an experiment to investigate how the thickness affects the insulating properties of newspaper.                                                             [6]

# STANDARD CIRCUIT DIAGRAM SYMBOLS

Electric circuit diagrams are drawn using standard **circuit symbols** for each of the different **components** that make up a circuit.

### Standard symbols

The symbols below are universally recognised. This saves time drawing or describing them in other ways.

Some symbols are quite similar. You will need to be able to recall and draw them accurately, preferably with a pencil and ruler.

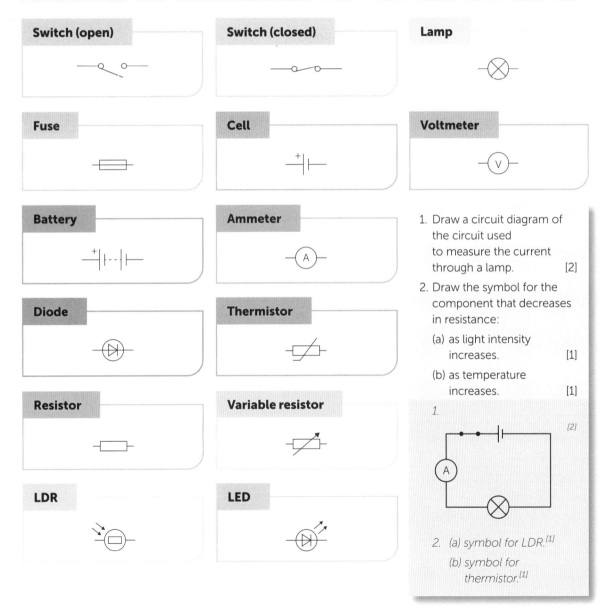

**Switch (open)**

**Switch (closed)**

**Lamp**

**Fuse**

**Cell**

**Voltmeter**

**Battery**

**Ammeter**

1. Draw a circuit diagram of the circuit used to measure the current through a lamp. [2]

2. Draw the symbol for the component that decreases in resistance:

   (a) as light intensity increases. [1]

   (b) as temperature increases. [1]

**Diode**

**Thermistor**

**Resistor**

**Variable resistor**

1.

[2]

**LDR**

**LED**

2.  (a) symbol for LDR.[1]

(b) symbol for thermistor.[1]

# ELECTRICAL CHARGE AND CURRENT

Electric **current** is the **rate of flow of electrical charge**. In metal wires, this charge is carried by **electrons**.

### Calculating charge

The rate of flow of charge, or size of the current, can be measured by determining how much charge passes by a point in the circuit each second. The current in a component is measured using an **ammeter** –Ⓐ– connected in **series** with the component. The current has the same value anywhere in a single closed loop.

For an electric charge to flow, the circuit must be a **closed circuit**. This means it is an unbroken loop, so any switches must be closed (on).

The circuit also needs a source of **potential difference** so must include a cell –∣⊢,
battery –∣⋅∣⊢ or power supply.

The charge flow can be calculated using the equation:

> You need to be able to recall and apply this equation.

$$charge\ flow = current \times time$$

$$Q = I \times t$$

$Q$ = charge flow in coulombs, C

$I$ = current in amperes (amps), A

$t$ = time in seconds

> Remember: It is the charge that flows, not the current.

> Learn the symbols for the quantities $Q$ and $I$ and the units C and A. Make sure not to get these quantities and units confused.

1. Complete the sentence.
   One coulomb is the _____ that passes a point in a circuit when there is a _____ of one ampere for one second. [2]

2. The current in a lamp is 0.7 A.
   Calculate the charge, in C, that flows through the lamp in 1.5 minutes. [3]

3. A charge of 0.12 C flows through a resistor in 86 s. Calculate the current, in A, in the resistor. [3]

4. There is a current of 0.80 A in a circuit. Calculate the time taken, in s, for 32 C of charge to flow past a point in the circuit. [3]

1. Charge[1], current.[1]

2. $t = 1.5 \times 60 = 90\ s$[1]

   $Q = It$
   $= 0.7 \times 90$[1]
   $= 63\ C$[1]

3. $Q = It$
   $0.12 = I \times 86$[1]
   $I = \frac{0.12}{86}$[1]
   $= 0.0014\ A$[1]

4. $Q = It$
   $32 = 0.80 \times t$[1]
   $t = \frac{32}{0.80}$[1]
   $= 40\ s$[1]

# CURRENT, RESISTANCE AND POTENTIAL DIFFERENCE

**Resistance** is a measure of how hard it is for charge to flow through a component or material.

## Potential difference and resistance

Materials with extremely high resistance are **electrical insulators**. The current in a component depends on the **potential difference** across the component and the resistance of the component. The potential difference across a component can be thought of as the difference in energy carried by electrons before and after they have flowed through a component. It is a measure of the electrical work done.

For a given potential difference, the current decreases as the resistance increases.

The greater the resistance of the component, the smaller the current for a given potential difference across the component.

## Calculating potential difference

Potential difference, current and resistance are linked by the equation:

potential difference =
current × resistance

$$V = I \times R$$

$V$ = potential difference in volts, V
$I$ = current in amperes (amps), A
$R$ = resistance in ohms, $\Omega$

You need to be able to recall and apply this equation.

Potential difference (pd) is also known as voltage. In an exam, the term 'potential difference' will be used but the correct use of either term will gain marks.

1. The resistance of a resistor is 250 $\Omega$. The current in the resistor is 1.2 A.
   Calculate the potential difference across the resistor, in volts. [2]

2. The potential difference across a component is 4.5 V. The resistance of the component is 2.5 $\Omega$.
   Calculate the current, in amps, in the component. [3]

3. The potential difference across a resistor is 30 V .The current in the resistor is 0.50 A .
   Calculate the resistance, in ohms, of the resistor. [3]

4. Explain why the resistance in a wire changes as the wire gets hotter. [3]

1. $V = IR$
   $= 1.2 \times 250$ [1]
   $= 300$ V [1]

2. $V = IR$
   $4.5 = I \times 2.5$ [1]
   $I = 4.5 / 2.5$ [1]
   $= 1.8$ A [1]

3. $V = IR$
   $30 = 0.50 \times R$ [1]
   $R = 30 / 0.50$ [1]
   $= 60 \Omega$ [1]

4. As the wire gets hotter the atomic vibrations are larger [1], so electrons collide with atoms more frequently [1] and the resistance increases [1].

# REQUIRED PRACTICAL 3 (15)

## Investigating resistance

This practical activity helps you develop your ability to use appropriate apparatus and circuit diagrams to measure electrical quantities.

### Investigating the effect of length on the resistance of a wire at a constant temperature

- Set up a circuit to measure the potential difference and current for a wire at different lengths along the wire.

- Calculate the resistance for the different lengths of wire using $R = V / I$

- Plot a graph of resistance against length of wire and draw a line of best fit.

- Make a conclusion about the relationship between resistance and length.

The investigation should lead to the conclusion that the longer the wire the higher the resistance.

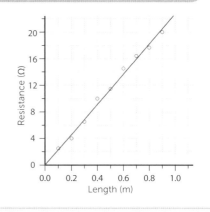

1. (a) Describe what the results in the graph show. [1]

    (b) Explain how changing the length of the wire affects the resistance. [2]

2. Give **two** ways to reduce errors caused by the heating effect of a current when investigating the resistance of a wire. [2]

3. At a wire length of 200.0 cm, the pd is measured at 0.53 V, 0.58 V and 0.54 V.

    (a) Calculate the mean potential difference (pd). [1]

    (b) The mean current is 0.22 A. Calculate the resistance of the wire at 200.0 cm. [3]

1. *(a) The resistance increases as the length of wire increases.[1]*

    *(b) Resistance is directly proportional to the length of the wire.[1] For a shorter wire, electrons do not have to travel so far through the wire so resistance is less.[1]*

2. *Use a low potential difference to keep current low and avoid heating the wire too much.[1] Only turn current on briefly to take the reading / allow the wire to cool between readings.[1]*

3. *(a) (0.53 + 0.58 + 0.54) / 3 = 0.55 V[1]*

    *(b) V  = IR*
    *0.55 = 0.22 × R[1]*
    *R   = 0.55 / 0.22[1]*
    *    = 2.5 Ω[1]*

To increase the reliability of the results:

Repeat the values and calculate a mean resistance for each length. Use smaller gaps between lengths.

**Sources of error:**

- The heating effect of the wire.
- Measuring the length between the crocodile clips consistently.

## Resistance of resistors arranged in series and parallel

- Use circuit diagrams to set up circuits connecting two **identical resistors**, $R_1$ and $R_2$, in series and then in parallel.
- Measure the potential difference across the resistors and the current in both circuits.
- Calculate the total resistance in the circuit.
- Compare the resistance of each arrangement of resistors.

**2 resistors in series**

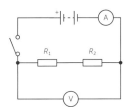

**2 resistors in parallel**

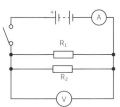

It is important to connect the components correctly. This makes sure the resistances can be compared accurately and safely, and also protects the components. There should be positive readings on both the ammeter and the voltmeter. If you see a negative value, swap the connections around. The smaller the **resolution** of the ammeter and voltmeter, the more decimal places the readings can be given to.

| Circuit | Potential difference (V) | Current (A) |
|---------|--------------------------|-------------|
| Series  | 6.4                      | 0.16        |
| Parallel | 5.9                     | 0.59        |

4. The table above shows the results for two identical resistors connected in series and in parallel.

Use the equation resistance = potential difference / current.

(a) Calculate the total resistance in the series circuit. [1]

(b) Calculate the total resistance in the parallel circuit. [1]

(c) Compare the resistance of the two arrangements. [2]

4. (a) Series: $V = IR$
   $6.4 = 0.16 \times R$
   $R = 6.4 / 0.16$
   $= 40\ \Omega$ [1]

   (b) Parallel: $V = IR$
   $5.9 = 0.59 \times R$
   $R = 5.9 / 0.59$
   $= 10\ \Omega$ [1]

   (c) The total resistance in the circuit is higher when the resistors are connected in series. [1]
   The resistance in series is four times higher than the resistance when they are connected in parallel. [1]

# RESISTORS

For some resistors, the value of $R$ (resistance) remains constant as the current changes. These resistors are **ohmic conductors**.

**Non-ohmic** resistors do not follow this relationship and the value of $R$ can change as the current changes. Filament lamps, diodes, thermistors and LDRs are non ohmic resistors. Their resistance changes with the current in the component.

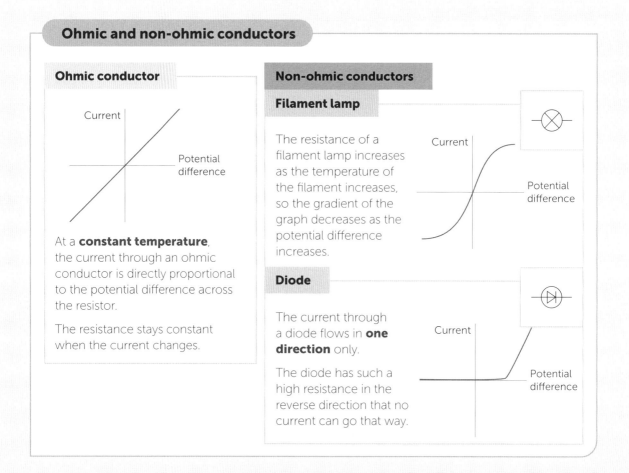

## Ohmic and non-ohmic conductors

### Ohmic conductor

At a **constant temperature**, the current through an ohmic conductor is directly proportional to the potential difference across the resistor.

The resistance stays constant when the current changes.

### Non-ohmic conductors

#### Filament lamp

The resistance of a filament lamp increases as the temperature of the filament increases, so the gradient of the graph decreases as the potential difference increases.

#### Diode

The current through a diode flows in **one direction** only.

The diode has such a high resistance in the reverse direction that no current can go that way.

Remember the basic circuit for measuring the resistance of a component:

- Ammeter in series with the component
- Voltmeter in parallel (across the component).

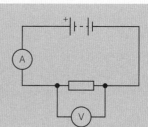

## Thermistor

The resistance of a thermistor decreases as the temperature increases.

These can be used as temperature sensors in fire alarm systems. Alarms or sprinklers are switched on when a particular temperature is exceeded. When the resistance becomes low enough, a current can flow in the circuit which is used to switch on the systems. They are also used in thermostats for switching on devices at required temperatures.

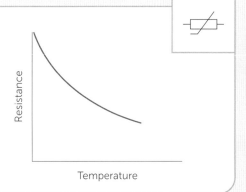

## LDR (Light Dependent Resistor)

The resistance of an LDR decreases as light intensity increases. In bright light, the resistance of an LDR is low so more current can flow through it. In the dark, or at low light levels, the resistance of an LDR is high, so much less current can flow through it.

This means LDRs can be used to detect light levels. They will switch on security and street lamps automatically at low light, and switch them off when light levels rise again.

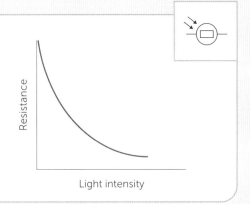

Explain the I-V graphs of component X and component Y. Include resistance and temperature in your answer. [6]

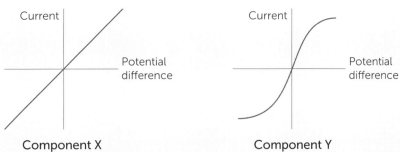

*Component X is an ohmic conductor[✓]. The current in it is directly proportional to the potential difference across it[✓]. As $V = I \times R$[✓], the resistance is constant whatever the potential difference is across it. The value of R can be found from the gradient of the graph[✓] which is I / V[✓] which equals 1 / R[✓]. For this relationship, the temperature of the component is constant.[✓]*

*Component Y is a non-ohmic resistor.[✓] The gradient of the graph decreases as potential difference increases[✓] which means that the resistance increases as the potential difference increases.[✓] This is caused by an increasing temperature[✓] from the heating effect of a current[✓].*

This is an extended response question that should be marked in accordance with the levels based mark scheme on page 170.

This practical activity helps you develop your ability to use appropriate circuits and circuit diagrams to measure current and potential difference for different components.

### Measuring current and potential difference

Investigate what happens to the current in (i) a **resistor**, (ii) a **filament lamp** and (iii) a **diode** when the potential difference across it is changed.

This is the circuit used for finding the characteristics of the diode.

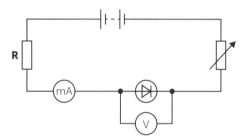

When testing the resistor and filament lamp, replace the diode and remove resistor *R*.

Use a normal ammeter instead of a milliammeter (mA) to measure current when testing these two components. This depends on the values of the current, and the resolution of the ammeters.

Set up the circuit for each component. Adjust the variable resistor and record several pairs of ammeter and voltmeter readings in a suitable table. Swap the connections on the power supply – connect the ammeter to the negative terminal and connect the component being tested to the positive terminal. This will give readings with **negative values**.

Plot a graph of current against potential difference for each component.

The I-V graphs plotted for each component (current against potential difference) should look similar to those on page 25. The shape of the graph is the *I-V* characteristic of the component.

---

1. The gradient of the graph for the resistor is a straight line that goes through the point (0, 0). Explain what this tells you about the resistor. [2]
2. Give reasons why the circuit to test the diode has an extra resistor in it and uses a milliammeter. [2]

1. *The resistor is an ohmic conductor[1] because the graph is directly proportional which means that the resistance is constant[1].*
2. *The extra resistor is used to protect the diode from high currents[1] and a milliammeter is used because the current values are low[1].*

# SERIES AND PARALLEL CIRCUITS

There are two ways of joining electrical components. A single continuous loop is a **series** circuit. A **parallel** circuit has components that are connected across each other. Circuits can include series and parallel parts.

## Comparing series and parallel circuits

Compare a cell and two lamps when connected in series and in parallel.

| | Series | Parallel |
|---|---|---|
| Circuit diagram | | |
| Potential difference | Total potential difference of the power supply is shared between the lamps. $V_{cell} = V_1 + V_2$ | Potential difference across each lamp is the same: $V_{cell} = V_1 = V_2$ |
| Current | Same through each lamp. $I_1 = I_2$ | Total current through whole circuit is the sum of the current through each lamp. $I_{total} = I_1 + I_2$ |
| Resistance | Total resistance is the sum of the resistance of each lamp. $R_{total} = R_1 + R_2$ The current passes through all the resistors in series so adding resistors decreases the current and increases the total resistance of the circuit | The total resistance is less than the resistance of the bulb with the smallest resistance in the circuit. The total current splits between resistors in parallel. Adding resistors means there are more paths for the current, so the total current increases and the total resistance decreases. |
| Lamps | Both off or both on. | Can be switched on and off separately. |

1. Look at the series circuit in the text above. $V_{cell}$ = 1.5 V, and the resistance of each lamp is 2 Ω.

    (a) Calculate the total resistance, in ohms, in the circuit. [1]

    (b) Calculate the current, in amperes, in the circuit. [3]

2. A series circuit consists of three identical lamps connected with a power supply of 12 V. Calculate the potential difference across each lamp. [1]

3. Look at the parallel circuit in the text above. The resistance of each lamp is 2 Ω. Describe the equivalent resistance in this circuit. [1]

1. (a) $R_{total} = R_1 + R_2$
      $= 2 + 2$
      $= 4 \, \Omega$

   (b) $V = IR$
       $6.0 = I \times 4$[1]
       $I = 6.0 / 4$[1] $= 1.5 \, A$[1]

2. 12 V / 3 lamps $= 4 \, V$[1]

3. It is less than 2 Ω[1]

# DIRECT AND ALTERNATING POTENTIAL DIFFERENCE

## dc and ac

Cells, batteries and solar cells have two terminals. The electrons always move from the negative terminal to the positive terminal. This means the current is always in the same direction and it is called a **direct current** (**dc**). (Note that conventional current is from positive to negative).

In an **alternating current** (**ac**) the direction of the movement of the electrons is constantly changing. This means that the potential difference is also constantly changing. The potential difference increases to a peak and then decreases to zero. It carries on decreasing to a negative 'peak' and then increases to zero. This cycle repeats continuously, 50 times each second in the UK.

## pd-time graphs to show direct and alternating potential difference

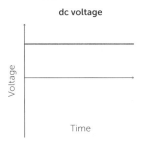

dc voltage

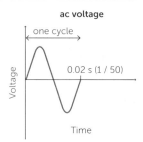

ac voltage

**dc**

The potential difference-time graph shows a straight line at a constant pd.

**ac**

The potential difference-time graph shows a curve alternating between positive and negative pd. The positive and negative values indicate the direction of the current.

1. Describe the electricity supply in the UK. Tick **two** boxes. [2]

   ☐ dc supply   ☐ Frequency of 50 Hz
   ☐ Frequency of 230 Hz   ☐ pd of 110 V
   ☐ pd of 230 V

2. Give **two** examples of devices in the home that use a dc supply. [2]

   *1. Frequency of 50 Hz[1] pd of 230 V[1]*

   *2. Any two devices that use a battery or a cell, not the mains supply. E.g. mobile phone, electric toothbrush, remote control, hand held vacuum cleaner.[1]*

# MAINS ELECTRICITY

## Mains plug

The cable from an electrical appliance to the mains consists of three separately insulated metal wires inside a tube of insulation. It is known as a **three-core cable**.

Each of the three wires inside is colour coded for identification. Inside the appliance plug, the wires are connected as shown. A fuse is also connected to the live wire. Then it is all secured with a cable grip.

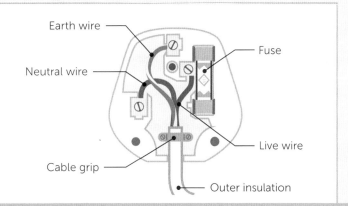

Earth wire

Neutral wire

Fuse

Cable grip

Live wire

Outer insulation

Electrical conductors carry a current e.g. metals.

Electrical insulators do not carry a current e.g. plastic and rubber.

| Wire | Colour | Potential difference with the earth | Purpose |
|------|--------|-------------------------------------|---------|
| Earth | Green and yellow stripes | 0 V | Safety wire. Only carries a current if there is a fault. |
| Live | Brown | About 230 V | Carries the alternating pd from the supply. Connects to a fuse. |
| Neutral | Blue | At or close to 0 V | Completes the circuit. |

## Safety features

A **fuse** ⊏═⊐ contains a wire that will melt if the current gets too high. If this happens, the appliance using the fuse will be at 0 V and not 230 V, so it is made safe.

If a person touches a live wire or appliance, the current travels through them to the ground and they get an electric shock. The **earth** wire is a safety wire to stop the actual appliance becoming live if the live wire touches a conducting part of the appliance, for example, the metal casing of a toaster.

1. Explain why different coloured insulation is used in a three-core cable.    [2]
2. Describe **two** dangers of using the mains supply.    [2]

    1. *To identify the different wires.[1] To connect wires correctly for safety / for the appliance to work.[1]*

    2. *Electric shock from contact with a live connection[1]. Fire caused by a too much current.[1]*

# POWER

The power transfer in any component or appliance is proportional to the potential difference across it and the current through it.

## Power, potential difference and current

Power, potential difference and current are linked by the equation:

$$\text{power} = \text{potential difference} \times \text{current}$$

$$P = V \times I$$

$P$ = power in watts, W

$V$ = potential difference in volts, V

$I$ = current in amperes (amps), A

Remember these definitions of power from **page 7**.

$$\text{power} = \frac{\text{energy transferred}}{\text{time}} = \frac{\text{work done}}{\text{time}}$$

You need to be able to recall and apply both these equations.

## Power, current and resistance

Power, current and resistance are linked by the equation:

$$\text{power} = (\text{current})^2 \times \text{resistance}$$

$$P = I^2 R$$

$P$ = power in watts, W

$I$ = current in amperes (amps), A

$R$ = resistance in ohms, $\Omega$

- As $V = I \times R$, substitute $I \times R$ for $V$ in the equation $P = V \times I$
- $\text{current}^2 = \text{current} \times \text{current}$

Large powers are given in kW.

1 kW = 1000 W

1. Convert 72.61 kW to W. Tick **one** box. [1]

   ☐ 0.07261 W    ☐ 7261 W    ☐ 72 610 W    ☐ 726 100 W

2. Calculate the power, in kW, of an electric iron with a current in it of 11.3 A and a potential difference across it of 230 V. [2]

3. There is a potential difference of 1.5 V across a 7.5 Ω resistor. The current through the resistor is 0.2 A. Show **two** different calculations to confirm that the power of the resistor is 0.3 W. [2]

4. **Higher Tier only:** A 5.0 Ω resistor has a power rating of 12.0W. Calculate the current, in amps, in the resistor. [3]

1.  *72 610 W*[1]

2.  *P = V × I*
    *= 230 × 11.3*[1]
    *= 2.60 kW*[1]

3.  *P   = V × I*
    *= 1.5 V × 0.2 A = 0.3 W*[1]
    *P   = I²R*
    *0.2 A × 0.2 A × 7.5 Ω = 0.3 W*[1]

4.  *P   = I²R*
    *12 = I² × 5*[1]
    $I \quad = \sqrt{\frac{12}{5}}^{[1]} = 1.5\ A^{[1]}$

# ENERGY TRANSFERS IN EVERYDAY APPLIANCES

Everyday electrical appliances are designed to transfer energy. They transfer energy from cells and batteries (dc supply) or from the mains (ac supply). Appliances with high power ratings need to use the mains supply.

An electrical appliance transfers more energy if it:
- has a **higher power** rating
- is switched on for **more time**

The energy is usually transferred:
- to make electric motors move (**kinetic** energy)
- for use in heating devices (**thermal** energy)

## Electrical work

When charge flows in a circuit, energy is transferred by electrical **work**. The amount of energy transferred by electrical work can be calculated using one of these equations.

**Energy transferred, power and time are linked by the equation:**

energy transferred = power × time

$$E = P \times t$$

$E$ = energy transferred in joules, J

$P$ = power in watts, W

$t$ = time in seconds, s

**Energy transferred, charge flow and potential difference are linked by the equation:**

energy transferred = charge flow × potential difference

$$E = Q \times V$$

$E$ = energy transferred in joules, J

$Q$ = charge flow in coulombs, C

$V$ = potential difference in volts, V

You need to be able to recall and apply these equations.

**Remember:** Electrical work done is equal to the energy transferred.

1. Give **one** example of a device that transfers useful energy by electrical work to:
   (a) kinetic energy [1]      (b) thermal energy. [1]

2. (a) A hairdryer has 3 heat settings and 2 speed settings. Explain why it has a power rating of 1650 W to 2000 W. [3]

   (b) Calculate the lowest amount of energy transferred, in kilojoules, when the hairdryer is used for 2 minutes. [3]

   (c) The potential difference across the hairdryer is 230 V. Calculate the charge flow, in coulombs, during the energy transfer in (b). [3]

   1. (a) *Any electrical device with a spinning motor e.g. washing machine, hairdryer, food mixer, fan.[1]*

      (b) *Any electrical device designed to get hot e.g. hairdryer, tumble dryer, iron.[1]*

   2. (a) *The power rating varies because the hairdryer transfers different amounts of energy[1] depending on the heat / speed settings.[1] Hotter / faster settings will have higher power ratings.[1]*

      (b) *$t = 2 \times 60 = 120 \ s$[1]*
          *$E = P \times t$*
          *$1650 \times 120$[1] $= 198 \ kJ$[1]*

      (c) *$E = Q \times V$*
          *$198 \ 000 = Q \times 230$[1]*
          *$Q = 198 \ 000 \ / \ 230$[1] $= 861 \ C$[1]*

# THE NATIONAL GRID

**Electrical power is transferred from power stations to consumers using the National Grid.**

The National Grid is the system of cables and transformers in the UK that links power stations to consumers.

**Step-up transformers** are used to increase the potential difference (pd) from the power stations. The pd is increased from about 25 kV to values up to 400 kV.

**Step-down transformers** are used to decrease the pd from the transmission cables to a much lower pd of 230 V for use in homes.

The current heating effect means that the current in the transmission cables heats up the cables, dissipating energy to the surroundings. This reduces the amount of useful energy transferred and makes the system less efficient.

The electrical power transferred by the cables is given by the equation:

$$power = pd \times current$$

So for the same power, when the pd is increased, the current decreases.

The electrical power dissipated by the cables is given by the equation:

$$power = (current)^2 \times resistance$$

So by reducing the current, and using wires with a low resistance, the electrical power dissipated by the current heating effect is minimised.

Explain why the National Grid system is an efficient way to transfer energy.    [6]

*The power dissipated in the transmission cables is given by the equation $P = I^2R$.[1] In the National Grid, a step up transformer[1] steps up the pd and reduces the current[1]. So, reducing the current, reduces the energy dissipated[1] by the current heating effect[1]. In addition, wires with low resistance are also used to help minimise the energy dissipated.[1]*

# STATIC CHARGE

**Static charge** is charge that builds up because it is not able to flow away to the surroundings. The electrons become static in movement. **Insulators** can become charged by **friction** when **electrons** are **transferred** to or from an **insulating material** (e.g. cloth, wool, glass, plastic, rubber).

## Electrons

It is only electrons that are transferred. Positive charges cannot move.

Electrons are negatively charged so:
- the material that gains electrons becomes negatively charged.
- the material that loses electrons is left with an equal positive charge.

When two electrically charged objects are brought close together, they exert a force of repulsion or attraction on each other. This is an example of a **non-contact force**. The closer the charged objects are to each other, the stronger the force of attraction or repulsion.

If the objects carry the **same** type of charge, **+/+** or **−/−** , they **repel**.
If the objects carry **opposite** types of charge, **+/−**, they **attract**.

## Sparks

**Sparks** occur when the electrons discharge to a conducting material or to the ground through the air. The object becomes discharged or earthed.

**Example:** A car passenger transfers electrons between their clothes and the car seat when they rub against each other. When the person touches the metal door (a conductor), the charge discharges and they experience a small electric shock.

1. Describe what happens when a polythene rod is rubbed with a cloth. [3]
2. Before aeroplanes are refuelled, they are connected to earth by a bonding line.
   Explain why the aeroplane is earthed before refuelling starts. [3]
3. A person becomes electrostatically charged and their hair stands out.
   Explain how this shows evidence of electrostatic forces. [2]

   1. *Electrons from the cloth are transferred to the polythene rod[1] this gives the rod a negative charge[1] and leaves the cloth with a positive charge[1].*
   2. *As the fuel flows through the pipe, friction causes a charge to build up.[1] The charge must be conducted to earth[1] to avoid a spark which could ignite the fuel vapour[1].*
   3. *The person's hair has become charged with the same type of charge (positive or negative)[1] and the person's hair is standing on end because there is a force of repulsion between charges of the same type[1].*

# ELECTRIC FIELDS

## Electric field around a charged object

An **electric field** is created around a charged object. The field is strongest close to the object. Another charged object placed inside the field will experience a **non-contact force**. Depending on the type of charges, the forces will be attractive or repulsive and the objects can move if they are free to do so.

- Imaginary field lines are used to represent an electric field.
- The field lines have arrows drawn from positive (+) to negative (−) showing the way a positive charge would be pushed in the field.
- A stronger field is represented by the field lines being drawn closer together. A weaker field is represented by the field lines being drawn further apart.
- The force experienced is stronger as the objects get closer and weaker as they move further apart.

The field lines spread out in a **radial field** from a single point.

This is a 2D diagram representing **isolated charged spheres**, in reality the field lines are all around it in 3D.

When a negatively charged electron is placed in the electric field around a positively charged nucleus, it is attracted to the positive charge. This idea helps to explain the non-contact force between objects. It also helps to explain sparking.

---

1. Define an electric field [1]
2. Draw the electric field lines for a charged sphere in isolation when it has:
   (a) a small positive charge. [1]
   (b) a large negative charge. [1]

    *1. An electric field is a region where an electric charge experiences a force.*[1]

    *2. (a)*         *[1]*     *(b)*         *[1]*

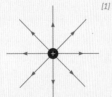

# EXAMINATION PRACTICE

01   A student measures the resistance of a lamp.

   01.1   Which **two** components are needed to measure the resistance? Tick **two** boxes.          [2]
   ☐ Ammeter   ☐ Battery   ☐ Voltmeter   ☐ Resistor   ☐ Switch

   01.2   The student uses a battery for a power supply.
   Draw the circuit diagram used to measure the resistance across a lamp.          [3]

   01.3   The current in the lamp is 0.5 A. The potential difference of the battery is 1.5 V.
   Calculate the resistance of the lamp.          [3]

   01.4   A charge of $3.0 \times 10^4$ C is transferred by a current of 250 mA.
   Calculate the time taken in seconds for this charge to be transferred.          [4]

02   This question is about understanding components.

   02.1   Name this component.          [1]

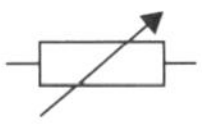

   02.2   Explain how this component works.          [2]

   02.3   Which **one** of these components is used to detect light? Tick **one** box.          [1]
   ☐ 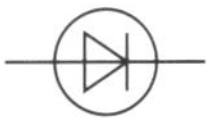          ☐ 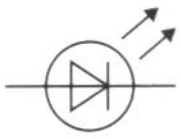          ☐ 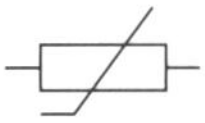          ☐

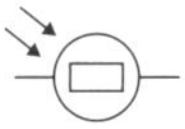

   02.4   Describe how this component is used to control circuits.          [2]

03   The diagram shows an electric circuit.

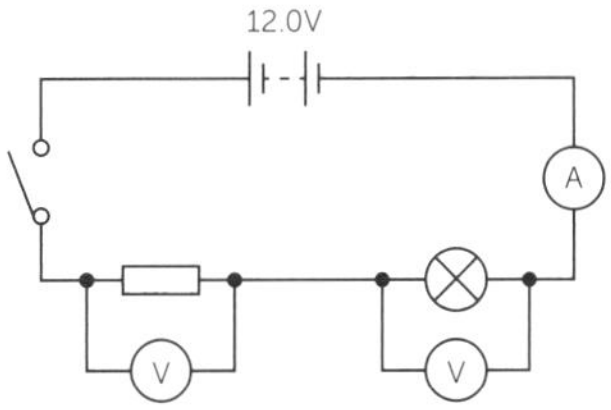

12.0V

The resistance of the lamp is 2.0 Ω. The ratio of the resistance of the resistor to the resistance of the lamp is 2:1.

   03.1   Calculate the equivalent resistance in the circuit.          [2]

   03.2   Calculate the reading on the ammeter.          [2]

   03.3   Calculate the potential difference across the:
   (a) resistor [1]     (b) lamp [1]

   03.4   The resistance of a wire is proportional to its length. Which graph represents this?
   Tick **one** box.          [1]

   A ☐ 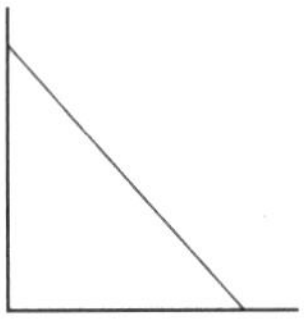          B ☐ 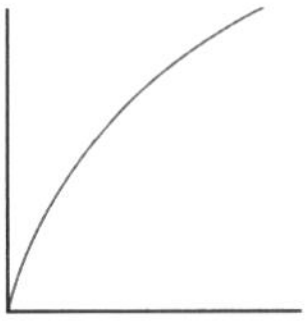          C ☐ 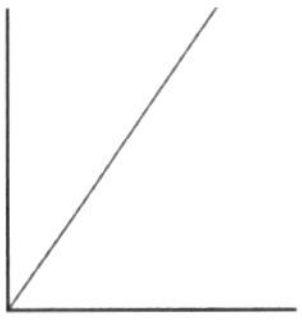          D ☐

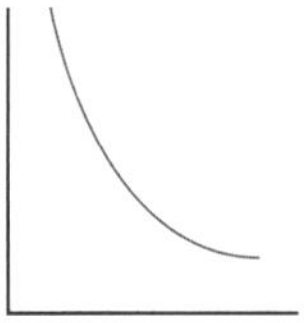

04 Describe the difference between direct and alternating potential difference. [2]

05 Explain the danger of the live wire touching a conducting part of an electrical appliance. [2]

06 Describe how transformers are used in the National Grid. [3]

07 An electric kettle with a power rating of 2.9 kW is connected to the 230 V mains supply.

    07.1 Calculate the current in the kettle. [3]

    07.2 A different kettle has a lower power rating and is connected to the same mains supply. Compare the current in both kettles. Tick **one** box. [1]

       ☐ The kettle with the higher power rating has a lower current

       ☐ Both kettles have the same current

       ☐ The kettle with the higher power rating has a higher current.

    07.3 Write the equation that links current, power and resistance. [1]

08 A microwave oven transfers 45 000 J in 70 s when it is connected to the 230 V mains supply.

    08.1 Calculate the power, in watts, of the microwave oven. [3]

    08.2 Calculate the charge flow through the microwave oven in 70 s. [3]

09 An acetate (plastic) rod loses electrons when it is rubbed with a cloth.

    09.1 What happens to the charge on both the acetate rod and the cloth? Tick **one** box. [1]

       ☐ Both the acetate rod and the cloth have a negative charge.

       ☐ The acetate rod has a negative charge and the cloth has a positive charge.

       ☐ The acetate rod has a positive charge and the cloth has a negative charge.

       ☐ Both the acetate rod and the cloth have a positive charge.

    09.2 The acetate rod is moved towards another charged acetate rod. Describe what happens. [2]

    09.3 The circle represents an isolated electron as a sphere.
Draw the electric field produced by the electron. [2]

    09.4 Explain how a person can receive a small electric shock when they reach out towards a door handle. [2]

# DENSITY OF MATERIALS

The **particle model** describes the arrangement of particles in the three different **states of matter**. It is used to predict and explain the behaviour of **solids**, **liquids** and **gases** in different conditions and **densities**.

## The particle model

The **particle model** represents the arrangement and movement of particles in the three different states. The particles can be atoms, ions or molecules, but all of them are shown as small, solid spheres.

| | Solid | Liquid | Gas |
|---|---|---|---|
| Arrangement of particles | Regular | Random | Random |
| Relative distance between particles | Close | Close | Far apart |
| Motion of particles | Vibrate about fixed positions | Move around each other | Move quickly in all directions |
| Particle diagram | | | |

## Calculating density

The density of a material is the mass of a substance per unit volume. **Volume** is how much space a material occupies. For two materials with the same volume, the one with the larger mass will have the higher density. Density, mass and volume are linked by the equation:

$$\text{density} = \frac{\text{mass}}{\text{volume}} \qquad \rho = \frac{m}{V} \qquad \frac{m}{\rho \times V}$$

$\rho$ = density in kilograms per metre cubed, kg/m³.

$m$ = mass in kilograms, kg

$V$ = volume in metres cubed, m³

You need to be able to recall and apply this equation.

Density is often given in g/cm³
1 g/cm³ = 1,000 kg/m³

1. Explain the difference in density between the same mass of a substance when it is a solid, a liquid and a gas. Use the particle model. [6]

2. A solid has a mass of 36 g and a volume of 180 cm³. Calculate the density of the solid. [2]

1. The density of the same mass of substance will be highest for the solid, a little lower for the liquid and much less for the gas.[1] For the same mass of a substance, there are the same number of particles[1] and the higher the volume, the lower the density.[1] The particles in a solid are packed closely together so have the highest density.[1] When the same mass is in liquid form, there is a little more space between the particles.[1] However, when the substance is in gas form, the particles are spread much further apart so takes up more volume.[1]

2. $\rho = m/V$
   $= 36 / 180$[1] $= 0.2$ g/cm³.[1]

# REQUIRED PRACTICAL 5 (17)

## Determining the densities of liquids, regular solids and irregular solids

This practical activity helps you develop your ability to measure length, mass and volume accurately, and then determine densities.

### Density, mass and volume

To determine density, the mass and volume of a subject needs to be measured.

Use $\text{density} = \dfrac{\text{mass}}{\text{volume}}$

Mass is measured using an electronic balance. The mass of a solid can be measured directly by placing it onto a balance. A liquid can be put into a container with a known mass.

The mass of the liquid = (mass of liquid and container) − (mass of empty container)

Volume is measured in different ways depending on both the state and shape of the substance.

### Liquids

In a classroom laboratory, the most accurate container for measuring the volume of a liquid is a **measuring cylinder**. For accuracy, choose one with the smallest graduations that will contain the volume of liquid.

**Example:** For an approximate volume of 4 cm³, use a 5 cm³ cylinder with graduations of 0.1 cm³ instead of a 10 cm³ cylinder with graduations of 0.2 cm³.

**Liquid volume is also measured in ml.**
**(1 cm³ = 1 ml)**

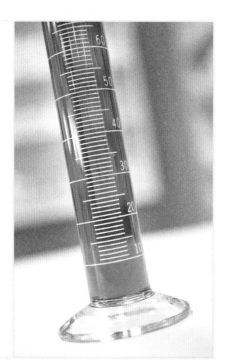

### Regular shaped solids

The volume of a **regular shape** can be found by measuring the quantities that are needed to calculate the volume from the mathematical formula for that shape. This usually includes **length**.

**Example:**
Volume of a cuboid = length × height × width

Measure the three lengths using a **ruler** or **Vernier callipers**.

## Irregular shaped solids

The volume of an irregular shape can be found by displacement. The solid is carefully lowered into a liquid, usually water. The solid will displace the same volume of the liquid and this volume can then be measured. The solid must be fully submerged to obtain its volume.

### 1. Using a eureka can

The can is filled up to the spout with water and the solid lowered in. The displaced water is collected in a measuring cylinder. The displaced water has the same volume as the irregular solid. Because a liquid takes the shape of a container, the volume can be easily measured.

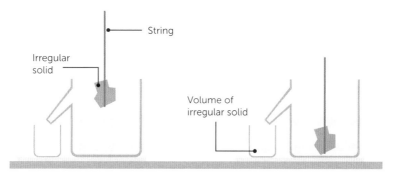

### 2. Direct measurement in a measuring cylinder

Record the volume of the water in the measuring cylinder. Carefully lower the solid into the water and record the new volume.  volume of object = (volume with object) − (volume without object)

What potential sources of error are there in the two methods described above?                    [3]

> *Not lowering in the object carefully so that water splashes out of the measuring cylinder.*[1]
> *Not reading the scale on the measuring cylinder at eye level.*[1]
> *Level of water in the eureka can is below the level of the spout when the object is lowered into it.*[1]

# CHANGES OF STATE

Changes of state are **physical changes**. This means they can be reversed. The substance can go back to having the original properties by doing the opposite process e.g. heating instead of cooling. This is not the case for a chemical change.

## Describing changes of state

Energy is transferred in to, or out of, the substance's thermal energy store by heating or cooling. When a substance changes from one state to another, the arrangement of, movement of, and distance between the particles changes. The **mass** of the substance is **conserved** when it changes state as the number of particles does not change. Even if the particles escape, (e.g. as a gas) the particles still exist.

**Boiling** and **vaporisation** are the same process as they both describe the change of state from a liquid to a gas. Evaporation is also the process of a liquid changing into a gas, but it is not the same as vaporisation. It takes place at temperatures below the boiling point of the liquid. In evaporation, the particles with the most energy escape from the surface of the liquid to form a vapour. A substance's vapour is a gas that has a temperature below the boiling point of the substance's liquid.

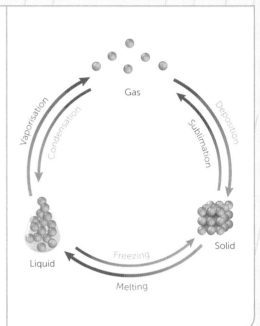

1. Explain what happens to the particles when a substance melts. [3]
2. Define sublimation. [1]
3. Droplets of water form on the inside of a window on a cold day.
   Explain why this happens. [3]

   1. *Energy is transferred to the particles by heating[1] and the particles gain thermal energy[1] so can move away from their fixed positions and start to move around each other as a liquid[1].*

   2. *Sublimation occurs when a substance changes state directly from a solid to a gas.[1]*

   3. *Energy is transferred from the water particles in the air[1] to the colder glass[1] and water changes from the gas state to the liquid state[1].*

# INTERNAL ENERGY

The energy stored in the particles of a system is called **internal energy**. This is the **total kinetic energy** and **potential energy** of all the particles (atoms and molecules) that make up the system.

## Energy is stored in particles

The **particle model** shows us that particles move in the liquid and gas states. This means they have a kinetic energy store. Particles also have a potential energy store due to any bonds and intermolecular forces between them.

The energy stored within the system changes when the system is heated or cooled.

- **Heating** increases the internal energy of the particles that make up the system.
- **Cooling** decreases the internal energy of the particles that make up the system.

The effect of this energy change when a system is heated or cooled either changes the temperature of the system, or produces a change of state.

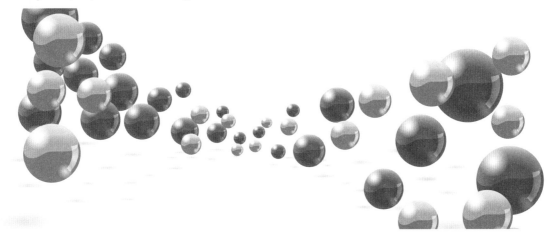

Remember: a system is an object or a group of objects.

1. Give **two** changes that can happen to a liquid when the internal energy is increased.    [2]
2. Describe what happens to particles in a gas when the temperature rises.    [2]
3. Which form of the same mass of a substance has the most internal energy? Tick **one** box.    [1]

        ☐ Gas        ☐ Liquid        ☐ Solid

1. *The temperature of the liquid increases[1] the liquid can change into a gas[1].*
2. *The particles gain kinetic energy[1] and move faster[1].*
3. *Gas.[1]*

# TEMPERATURE CHANGES IN A SYSTEM AND SPECIFIC HEAT CAPACITY

When the temperature of a system changes, the change in temperature depends on the mass of the substance heated, the type of material and the energy change in the system. Each substance has an individual specific heat capacity.

## Thermal energy

The change in thermal energy stored or released, as the temperature of a system varies, can be calculated using the equation:

change in thermal energy = mass × specific heat capacity × temperature change

$$\Delta E = m\,c\,\Delta\theta$$

$\Delta E$ = change in thermal energy in joules, J

$m$ = mass in kilograms, kg

$c$ = specific heat capacity in joules per kilogram per degree Celsius, J/kg °C

$\Delta\theta$ = temperature change in degrees Celsius, °C

> You need to be able to select this equation from the equation sheet and apply it.

Know the difference between:

- Specific heat capacity – the amount of energy required to raise the temperature of one kilogram of a substance by one degree Celsius.

- Specific latent heat – the amount of energy required to change the state of one kilogram of a substance at a constant temperature.

A brick has a mass of 0.345 kg. The specific heat capacity of brick is 840 J/kg °C.

(a) Calculate the increase in temperature when 6320 J is transferred to the brick. [2]

(b) How much energy is transferred from the brick as it cools back to the original temperature? [1]

(a) $\Delta E$   = $m\,c\,\Delta\theta$

    6320  = 0.345 × 840 × $\Delta\theta$

    $\Delta\theta$    = 6320 / (0.345 × 840)[1]

          = 21.8 °C.[1]

(b) 6320 J is transferred from the brick to the surroundings.[1]

# CHANGES OF STATE AND SPECIFIC LATENT HEAT

Energy is needed for a substance to change state from a solid to a liquid, and from a liquid to a gas. This is called the **latent heat**. The same amount of energy is released when the reverse change of state happens as the substance cools.

## Changes of state

During a change of state, the energy transferred to the substance changes the energy stored in the substance (the internal energy). The temperature of the substance stays constant and only starts to change when the change of state is complete.

- For a solid changing to a liquid this temperature will be the **melting point** of the substance.
- For a liquid changing to a gas, this temperature will be the **boiling point** of the substance.

## Specific latent heat

The specific latent heat of a substance is the amount of energy required to change the state of one kilogram of the substance with **no change** in temperature.

The change in thermal energy, stored or released as a substance changes state, can be calculated using the equation:

$$\text{energy for a change of state} = \text{mass} \times \text{specific latent heat}$$
$$E = m \times L$$

$E$ = energy in joules, J

$m$ = mass in kilograms, kg

$L$ = specific latent heat in joules per kilogram, J/kg

> You need to be able to select this equation from the equation sheet and apply it.

| Specific latent heat of fusion | Specific latent heat of vaporisation |
|---|---|
| for the change of state from solid to liquid | for the change of state from liquid to gas |

1. Define:
   (a) Specific latent heat of fusion. [1]

   (b) Specific latent heat of vaporisation. [1]

2. The specific latent heat of vaporisation of nitrogen is 199 200 J/kg.
   Calculate the amount of energy needed to change 0.35 kg of liquid nitrogen into a gas. [2]

   1. *(a) Specific latent heat of fusion is the energy required to change 1 kg of a solid to a liquid[1]*

      *(b) Specific latent heat of vaporisation is the energy required to change 1 kg of a liquid to a gas.[1]*

   2. *E  = m × L*
      *= 0.35 × 199 200[1] = 69 720 J[1]*

# STATE CHANGES IN HEATING AND COOLING CURVES

## Heating curve

This **heating curve** shows what happens to the temperature of a substance over time when a solid is heated, changes state to a liquid, and then to a gas. In between the changes of state, the substance is heating up in whatever state it is in.

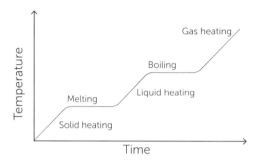

The melting point and boiling point can be found by reading the temperature from the graph when the line is horizontal.

## Cooling curve

A **cooling curve** is a graph showing what happens to the temperature of a substance over time when a substance is cooled. The changes of state will be gas to a liquid, and then liquid to a solid.

Remember: During a change of state, the temperature stays constant, so the line is horizontal.

The temperature at which a substance changes from a gas to a liquid is usually still called the boiling point rather than the condensing point. The temperature at which a substance changes from a liquid to a solid is the same temperature as the melting point, but this is sometimes called the freezing point too.

What is represented by the letters A, B, C, D, E, $T_1$ and $T_2$ shown on this cooling curve?                [4]

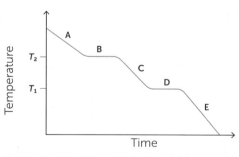

A:  A cooling gas; C: A cooling liquid; E: A cooling solid.[1]

B:  Changing state from a gas to a liquid (condensing).[1]

D:  Changing state from a liquid to a solid (freezing).[1]

$T_1$ is the melting point and $T_2$ is the boiling point.[1]

# PARTICLE MOTION IN GASES

The particles of a gas move about randomly and continuously, being free to fill up the space of a container. This means they collide with each other and the walls of the container they are in. The collisions exert a force on the container.

## Temperature and volume

The **average kinetic energy** of gas particles increases as the temperature of a gas increases.

This can be shown in a frictionless piston at different temperatures. The volume can change because the piston is frictionless. As the gas particles inside the piston are heated, the average kinetic energy increases; the particles move faster, and the piston expands. The opposite happens when the gas particles are cooled.

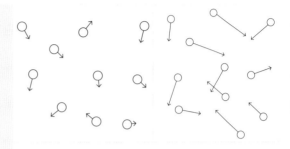

Gas particles are usually molecules. The movement of the gas particles is random.

## Temperature and pressure

However, when a gas is kept at a **constant volume**, increasing the temperature of the gas still increases the movement of the particles, but they cannot spread out. This means the pressure exerted by the gas increases.

The pressure increases when temperature increases because:

- the particles are moving faster so the force of each collision increases
- there are more particle collisions each second.

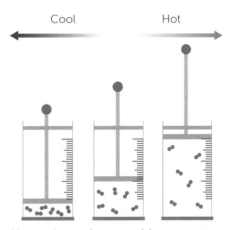

How volume changes with temperature

---

1. What happens to the pressure of a gas in a closed container when the temperature is decreased? Tick **one** box.                                                                                   [1]

   &#9744; Decreases      &#9744; Stays the same      &#9744; Increases

2. Explain how the motion of the particles in a fixed volume of gas is related to its temperature and pressure.                                                                                         [3]

   *1. Decreases.[1]*

   *2. The higher the temperature of the particles, the faster they move[1] and the more frequently they collide with each other and the container walls, increasing the force on the container walls[1]. This means that the pressure of the gas increases as the temperature increases.[1]*

---

# PRESSURE IN GASES

Gas particles create pressure which produces a net force at right angles to the wall of the container they are in.

## Calculating pressure

**Pressure** is the amount of force per unit area. It is measured in newtons per square metre ($N/m^2$) as well as in pascals (Pa).

Changing the pressure can cause a gas to be **compressed** or to **expand**.

For a **fixed mass of gas** held at a **constant temperature**, the pressure and volume of a gas are linked by the equation:

$$pV = constant$$

$$pressure \times volume = constant$$

$p$ = pressure in pascals, Pa

$V$ = volume in metres cubed, $m^3$

You need to be able to select this equation from the equation sheet and apply it.

Pressure is inversely proportional to volume so when one goes up, the other goes down.

The diagram shows how the pressure decreases as a piston moves. The volume increases and the fixed mass of gas can spread out further.

1. A piston contains 0.73 $m^3$ of a trapped gas at 1000 Pa. The temperature stays constant. The pressure is increased to 1200 Pa. Calculate the volume of the gas at 1200 Pa. [3]

2. A gas is held in a container at a constant temperature. The volume of the container is increased.
   Use the particle model to explain how this leads to a decrease in pressure. [3]

1. $p_1V_1 = p_2V_2$
   $1000 \times 0.73 = 1200 \times V_2$ [1]
   so $V_2 = 730 / 1200$ [1] $= 0.61 m^3$ [1]

2. *When the volume is increased, the same number of particles have more space to move around.* [1] *This reduces the frequency of the collisions with each other and the container walls* [1] *so the force is reduced, and the pressure decreases* [1].

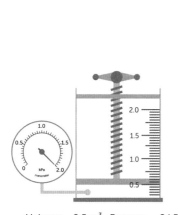

Volume = 0.5 $m^3$  Pressure = 2 kPa

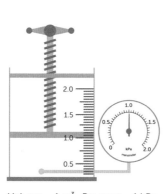

Volume = 1 $m^3$  Pressure = 1 kPa

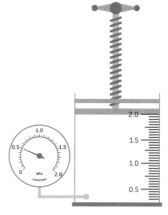

Volume = 2 $m^3$  Pressure = 0.5 kPa

# INCREASING THE PRESSURE OF A GAS

**Work** is the transfer of energy by a force.

Doing work on a gas increases the internal energy (the total kinetic and potential energy of the particles) of the gas. This can cause the temperature of the gas to increase because the temperature of a gas is related to the average kinetic energy of the particles.

Applying a force to compress a gas increases the pressure of the gas.

## Doing work on a gas

A person presses a piston down on a column of gas contained within the piston. The force moves the piston, and the gas is compressed. Work is done on the gas which means there is an energy transfer. Mechanical work has transferred energy from the person's chemical energy store to the internal energy store of the gas.

The amount of work done, or energy transferred, depends on the magnitude of the force and the distance the piston moves.

A bicycle pump is an example of a piston. A force is applied to compress a gas.

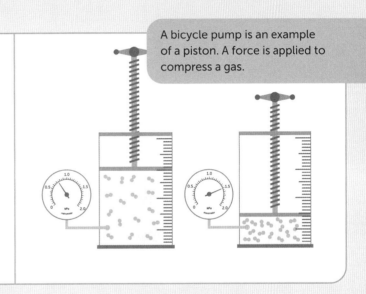

Explain why a bicycle pump gets warm when it is used to inflate a tyre.    [3]

*Work is done to compress the air in the bike pump[1] so energy is transferred to the air. This increases the internal energy of the air particles[1] so the temperature of the air increases[1]. Or:*

*The volume of the air in the pump decreases[1] so the pressure increases[1]. This increases the internal energy of the gas particles so the temperature increases.[1]*

# EXAMINATION PRACTICE

01   The density of a material is related to its mass and volume.
    01.1   Convert 736.4 g to kg. [1]
    01.2   763.4 g of a liquid has a volume of 50.0 cm$^3$. Calculate the density of the liquid in g/cm$^3$. [2]
    01.3   Calculate the volume of a cube with a side length of 0.02m. [2]
    01.4   A sphere has a density of 5400 kg/m$^3$. Calculate the mass of the sphere in g. [4]

02   This question is about the particle model.
    02.1   Use the particle model to describe what happens to the particles in a substance when it freezes. [3]
    02.2   Define internal energy. [1]

03   This question is about changes of state.
    03.1   Define specific latent heat of fusion. [1]
    03.2   The specific latent heat of fusion for water is 334 000 J/kg.
       Calculate the amount of energy released when 0.250 kg of water freezes. [2]
    03.3   Sketch a cooling curve to show a liquid freezing. [2]

**Higher only**

04   A 0.5 kg block of solid gold has a temperature of 1000 °C.
Calculate the amount of energy needed to change this block into liquid gold at 1064 °C. [5]
The melting point of gold is 1064 °C.
The specific heat capacity of gold is 130 J/kg °C. The specific latent heat of fusion for gold is 63 000 J/kg.

05   Complete the sentence. Tick **one** box. [1]
When the temperature of a gas, held at constant volume increases, the pressure exerted by the gas...
☐ Increases.
☐ Stays the same.
☐ Decreases.

**Physics only**

06    A piston contains 0.11 m³ of a trapped gas. The gas is compressed to a volume of 0.05 m³ at a pressure of 250 kPa.

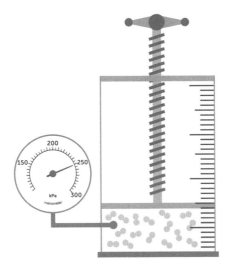

06.1   Calculate the initial pressure of the gas. The temperature stays constant.          [3]

**Physics (Higher) only**

06.2   Explain what happens to the temperature of the gas when it is compressed rapidly in the piston.          [3]

# THE STRUCTURE OF AN ATOM

The basic structure of an atom is a **positively charged nucleus** surrounded by **negatively charged electrons**.

## Atomic particles

Most of the mass of an atom is concentrated in the **nucleus** which contains its **protons** and **neutrons**.

- Protons are positively charged and have the same mass as a neutron.
- Neutrons have no charge so are neutral.

**Electrons** are about 1/2000 of the mass of a proton or a neutron and are tiny in comparison. Electrons orbit the nucleus in **energy levels**. Each energy level is at a fixed distance from the nucleus. Electrons can move from one energy level to another when electromagnetic radiation is absorbed or emitted.

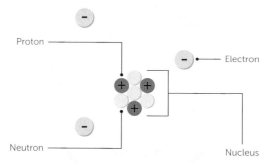

Proton

Electron

Neutron

Nucleus

Atom structure

The radius of an atom is about $1 \times 10^{-10}$ metres. Most of an atom is empty space between the outer electron energy level and the centre of the nucleus.

1. How many times bigger is the radius of an atom than the radius of the nucleus?
   Tick **one** box.                                                                                          [1]
   Approximately:
   ☐ 100    ☐ 1000    ☐ 10 000    ☐ more than 10 000

2. Describe what can happen in an atom when electromagnetic radiation is:
   (a) Absorbed by an atom.                                                                                   [1]
   (b) Emitted by an atom.                                                                                    [1]

   *1.  More than 10 000.[1]*

   *2  (a)  An electron moves into a higher energy level / further away from the nucleus.[1]*

   *(b) An electron moves into a lower energy level / closer to the nucleus.[1]*

# MASS NUMBER, ATOMIC NUMBER AND ISOTOPES

**The atomic number is the number of protons in the nucleus of an atom of an element.**

Atoms have zero net electrical charge. The number of positive protons in the nucleus equals the number of negative electrons in the energy levels of an atom.

All atoms of a particular element have the same atomic number. This means they all have the same number of protons and electrons.

Each element has a symbol. An atom is represented like this:

The **mass number** is the total number of **protons + neutrons** in an atom.

**Isotopes** are atoms of the same element that have **different numbers of neutrons**.

Mass number ———• 23

## Na

Atomic number ———• 11

### Ions

An atom forms a positive **ion** when it loses one or more electrons from the outer energy level.

**Isotopes of hydrogen**

Hydrogen-1
- 1 Electron
- 1 Proton
- 0 Neutrons

mass number = 1 + 0 = 1

Hydrogen-2
- 1 Electron
- 1 Proton
- 1 Neutron

mass number = 1 + 1 = 2

Hydrogen-1
- 1 Electron
- 1 Proton
- 2 Neutrons

mass number = 1 + 2 = 3

1. Describe what an atom of lithium $^{7}_{3}$Li is made up of.                                   [3]
2. Complete the sentence.                                                                          [2]

   Isotopes are atoms of the same element which have the same _____
   and a different _____.

   1. *3 protons and 4 neutrons[1] in the nucleus[1] with 3 electrons in energy levels around the nucleus[1].*
   2. *Isotopes are atoms of the same element which have the same atomic number / number of protons[1] and a different mass number / number of neutrons[1].*

# THE DEVELOPMENT OF THE MODEL OF THE ATOM

The **atomic model** has changed over time because of new experimental evidence.

## Atomic theories

In the early 19th century, atoms were imagined as tiny, solid spheres. The discovery of the **electron** by J.J. Thomson in 1897 led to his **plum pudding model**. The plum pudding model suggested that an atom was a ball of positive charge with negative electrons embedded in it. This model was disproved by a series of results from Rutherford's **alpha particle scattering experiment**.

In this experiment, beams of alpha particles (tiny positively charged particles) were aimed at thin gold foil. The results led to the **nuclear model**. Shortly afterwards, Niels Bohr carried out theoretical calculations showing that electrons orbit the nucleus at set distances. Observations from experiments supported his **electron shell model** and also showed the existence of positively charged **protons**. About 20 years later, James Chadwick demonstrated the existence of **neutrons**.

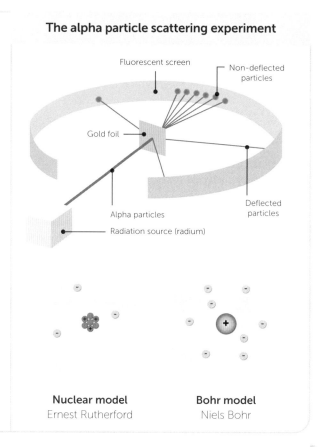

**The alpha particle scattering experiment**

Fluorescent screen · Non-deflected particles · Gold foil · Deflected particles · Alpha particles · Radiation source (radium)

Nuclear model
Ernest Rutherford

Bohr model
Niels Bohr

1. Compare the plum pudding and nuclear models of the atom. [3]
2. Explain why some scientific models have changed over time. [3]
3. Describe how the alpha particle scattering experiment led to the replacement of the plum pudding model by the nuclear model. [4]

   1. *Both models have negatively charged electrons.[1] These are embedded in a sphere of positive charge in the plum pudding model[1] but surround a positively charged nucleus in the nuclear model[1].*

   2. *Scientists obtain new experimental evidence[1] that does not support the current model[1]. If other respected scientists agree with the new findings the model will be changed.[1]*

   3. *All of the fast moving, positively charged alpha particles would have passed straight through the gold foil if the plum pudding model was correct.[1] However, some alpha particles were deflected[1] and a few were repelled straight back[1]. This led to the nuclear model idea that the atomic mass is concentrated in a positively charged centre.[1]*

# RADIOACTIVE DECAY AND NUCLEAR RADIATION

**Nuclear radiation comes from unstable atomic nuclei.**

## Radioactive decay

Some atomic nuclei are unstable because they have too many or too few neutrons.

**Radioactive decay** is a **random** process that occurs when an unstable atomic nucleus (**radioactive isotope**) gives out **nuclear radiation** to become more stable.

**Activity** is the rate at which a source of unstable nuclei decays. It is measured in **becquerel** (**Bq**). 1 Bq equals one nuclear decay per second.

**Count-rate** is the number of decays recorded each second by a detector.

## Types of nuclear radiation

| Name | Symbol | Composition | Range in air | Ionising power |
|------|--------|-------------|--------------|----------------|
| Alpha particle | α | A helium nucleus - two neutrons and two protons | < 5 cm | High |
| Beta particle | β | A high speed electron, ejected from the nucleus as a **neutron** turns into a proton | ~ 1 m | Low |
| Gamma ray | γ | Electromagnetic radiation which comes from the nucleus | Large distances | Very low |

A neutron (**n**) can also be emitted as a form of nuclear radiation.

The **ionising power** is the ability of the radiation to change an uncharged atom into a charged ion.

These different properties determine the choice of the radiation source when using radiation.

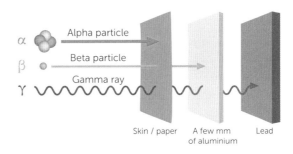

1. Describe how nuclear radiation can be detected and measured. [2]
2. The thickness of paper manufactured in a factory is monitored using a source of beta particles. Explain why sources of alpha particles and gamma rays are not used. [3]

   1. *A Geiger-Muller (GM) tube[1] measures the count-rate when connected to a ratemeter.[1] Or: Photographic film[1] turns foggy in the presence of radiation[1].*

   2. *Alpha particles cannot pass through the paper.[1] Gamma rays pass straight through so would not detect small changes in thickness.[1] Increasing the thickness of paper reduces the count rate of beta radiation so it is the most suitable source.[1]*

# NUCLEAR EQUATIONS

## Nuclear equations are used to represent radioactive decay.

Alpha particles and beta particles are emitted when unstable nuclei decay. This can cause a change in the charge or the mass of the nucleus which emits the particle.

Balanced **nuclear equations** can be written to show what happens when alpha or beta decay occurs.

An **alpha particle** is a helium nucleus so it is written as: $^{4}_{2}\text{He}$.

It has a mass number of 4 and an atomic number of 2. As there are no electrons to balance the protons, the alpha particle has a charge of +2.

### Writing nuclear equations

$$^{235}_{92}\text{U} \longrightarrow {}^{231}_{90}\text{Th} + {}^{4}_{2}\text{He}$$

$$231 + 4 = 235$$
$$90 + 2 = 92$$

A **beta particle** is a fast moving electron written as: $^{0}_{-1}e$
It has a mass number of 0 and a charge of −1.

$$^{137}_{55}\text{Cs} \longrightarrow {}^{137}_{56}\text{Ba} + {}^{0}_{-1}e$$

$$137 + 0 = 137$$
$$56 + -1 = 55$$

1. What happens to the mass number and atomic number of an unstable nucleus when it undergoes alpha decay? [2]

2. Explain why the emission of a gamma ray does not cause the mass or the charge of the nucleus to change. [2]

3. Balance this nuclear equation. [2]

$$^{222}_{86}\text{Rn} \longrightarrow \underline{\hspace{1cm}} \text{Po} + {}^{4}_{2}\text{He}$$

The atomic numbers and mass numbers must be balanced on both sides.

1. *The mass number decreases by 4[1] and the atomic number decreases by 2[1].*

2. *Gamma rays are electromagnetic radiation so have no mass[1] and no charge[1].*

3. $^{218}_{84}\text{Po}$, 218[1] 84[1]

# HALF-LIVES AND THE RANDOM NATURE OF RADIOACTIVE DECAY

**Radioactive decay is a random process.**

## Half-life

**Radioactive decay** is a random process because it is impossible to tell when a particular nucleus in a sample of a radioactive isotope will decay. However, it is possible to predict how many of the nuclei will decay in a certain amount of time.

The **half-life** of a radioactive isotope is:

- the **time** it takes for half of the unstable nuclei to decay.
- the **time** it takes for the count rate, or activity, to halve.

The half-life is different for each **radioactive isotope**.

## Calculating half-life

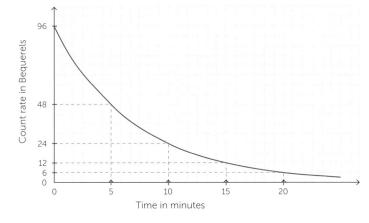

Draw a horizontal line from half of the initial activity level until it touches the graph line. The time given at the vertical line down will be the half-life. This should be the same whatever the starting point on the graph.

The graph shows that the count rate halves every 5 minutes. This is the case whatever the count rate is. The half-life of this radioactive isotope is therefore 5 minutes.

### Higher Tier only

The amount, activity or count rate of a radioactive isotope after a known number of half-lives, can also be calculated.

The **net decline** expressed as a ratio is **activity after *n* half-lives : initial activity** where *n* represents a number of half-lives.

1. Calculate the half-life of a sample when the measured count rate drops from 1600 to 200 in 6 days. [3]
2. **Higher only:** The half-life of dubnium-270 is 15 hours. Calculate the net decline after 45 hours. [3]

   1. *Half of 1600 is 800, half of 800 is 400, half of 400 is 200.*[1] *So, it takes 3 half-lives to drop from 1600 to 800.*[1] *3 half-lives are 6 days, so the half-life is 6 / 3 = 2 days.*[1]
   2. *45 hours is 3 half-lives.*[1] *½ x ½ x ½ = 1/8.*[1] *Ratio = 1:8.*[1]

# RADIOACTIVE CONTAMINATION

Exposure to radioactive materials can be harmful. Measures need to be taken to minimise the risk. There are two types of exposure to consider.

## Contamination and irradiation

### Radioactive contamination

**Contamination** involves unwanted direct contact with materials containing radioactive atoms. The exposed object also becomes radioactive. It is the decay of the radioactive atoms that causes the hazard. The level of danger is determined by the type of radiation emitted, its ionising power, and its ability to penetrate air and other materials.

Compare the hazards associated with radioactive contamination and irradiation.          [6]

*Contamination is the unwanted presence of radioactive particles on other objects after they and a radioactive material have come into direct contact with each other.[√] The object then becomes radioactive as well so will emit radiation[√] until all the particles have decayed[√]. The longer the half-life of the radioactive particles, the longer this takes.[√] Irradiation occurs when an object is exposed to a source.[√] The greater the distance from the source, the weaker the irradiation.[√] The shorter the time spent near the source, the lower the irradiation.[√] Removing the object from the source stops the irradiation.[√] The type of radiation affects the damage it can cause[√]. The more ionising the radiation, the more tissue damage.[√] Contamination with alpha particles inside a body is particularly dangerous[√] because alpha particles are highly ionising[√]. Being exposed to alpha particles outside the body is less harmful as they cannot penetrate skin.[√] Beta particles and gamma rays can penetrate skin.[√] Wearing protective clothing can reduce the risks from both contamination and irradiation.[√]*

This question should be marked with reference to the levels of response guidance on page 170.

## Irradiation

**Irradiation** is the process of exposing an object to nuclear radiation. The irradiated object does not become radioactive. For example, fresh fruit is irradiated to destroy bacteria on it and preserve the fruit for longer. The fruit does not become radioactive.

Humans are often irradiated deliberately for medical reasons with a controlled dose of radiation.

**Suitable precautions** must be taken to protect against any hazard. This includes wearing protective clothing, minimising unnecessary exposure and handling radioactive materials with tongs. Using tongs prevents contamination by stopping hands touching the radioactive material. They also reduce the rate of irradiation by increasing the distance of the hands from the radioactive source.

The effect of radiation on humans is studied globally. It is important to publish the findings so that other scientists can rigorously check them and confirm or challenge the evidence. This process is known as **peer review**.

# BACKGROUND RADIATION AND HALF-LIVES

## Sources of background radiation

**Background radiation** is around us all the time. It comes from **natural sources** and **non-natural** or **man-made sources**.

Some rocks naturally emit radon gas which is radioactive. Background radiation also comes from some food and drink. The Earth is continuously exposed to cosmic rays from space. Some medical uses, fallout from nuclear accidents and nuclear weapons testing all contribute to man-made background radiation.

### Background radiation sources in the UK

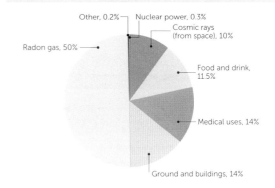

Other, 0.2%
Nuclear power, 0.3%
Cosmic rays (from space), 10%
Radon gas, 50%
Food and drink, 11.5%
Medical uses, 14%
Ground and buildings, 14%

1. Radiation dose is measured in sieverts (Sv).
   What is 27.3 Sv in millisieverts (mSv)? Tick **one** box. [1]

   ☐ 0.0273    ☐ 0.273    ☐ 2730    ☐ 27300

2. Give **two** factors that may affect a person's exposure to background radiation.
   Include an example for each. [2]

   *1. 27 300.[1] (1 Sv = 1000 mSv)*

   *2. Location – e.g. live near rocks with a high source of radon.[1]*
   *Occupation – e.g. a pilot has more exposure to cosmic rays.[1]*

## Different half-lives of radioactive isotopes

The half-life of a radioactive isotope is the time it takes for half of the unstable nuclei to decay

The half-life values of radioactive isotopes vary from a fraction of a second, to many millions of years. The longer the half-life is, the longer the radioactive isotopic decay will continue, and therefore produce a hazard for longer. This needs to be considered when choosing sources for different uses.

| Radioactive isotope | Half-life (seconds) |
|---|---|
| thorium-219 | $1.05 \times 10^{-6}$ |
| plutonium-228 | 1.1 |
| cobalt-60 | $1.66 \times 10^8$ (5 years, 3 months, 5 days) |

3. Look at the table of isotopes opposite.

   (a) Explain which of the isotopes given could be the most hazardous. [2]

   (b) What else do you need to consider when deciding on which source is the most hazardous? [1]

   *(a) Cobalt-60 could be the most hazardous[1] as is it has the longest half-life[1].*

   *(b) The activity of the source.[1]*

# USES OF NUCLEAR RADIATION

There are many uses of nuclear radiation.

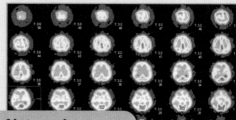

### Diagnostic medicine and treatments

The half-life and toxicity of an isotope used for medical reasons needs to be considered, as well as the type of radiation emitted in terms of ionisation and penetration.

### Medical contamination for the exploration of internal organs

Radioactive sources are used as **tracers** to look at soft tissues and to find areas of blockage. Sources emitting gamma rays can be injected into the body and detected by medical imaging processes.

This involves using non-poisonous isotopes with half-lives long enough for the isotope to produce enough measurements, but short enough so that the activity decays to safe levels quickly and the exposure of the patient to ionising radiation is minimised. They are usually emitters of gamma rays.

### Medical irradiation for the control or destruction of unwanted tissue

Nuclear radiation can damage or destroy cells which can be controlled for medical advantage.

Gamma rays can be used to kill cells in cancerous tumours deep inside the body. The dose is maximised by targeting the beam at different angles, and minimising damage to healthy tissue.

1. Suggest why tracers are usually emitters of gamma rays. [2]
2. Give **one** medical use of nuclear radiation other than in the body. [1]
3. Give **three** precautions to take when using radioactive sources. [3]

   1. *The detectors are outside the body[1] and alpha and beta particles would be stopped by the body and not detected by the detectors[1].*
   2. *Sterilisation of medical instruments.[1]*
   3. *Any **three** from: monitor exposure with a detector badge / avoid contact with skin / avoid breathing in / limit the exposure time / handle with tongs / keep sources in a lead box / wear protective clothing.[3]*

# NUCLEAR FISSION

**Nuclear fission** happens when a large unstable nucleus is split into two smaller nuclei usually by absorbing a neutron.

## Process of nuclear fission

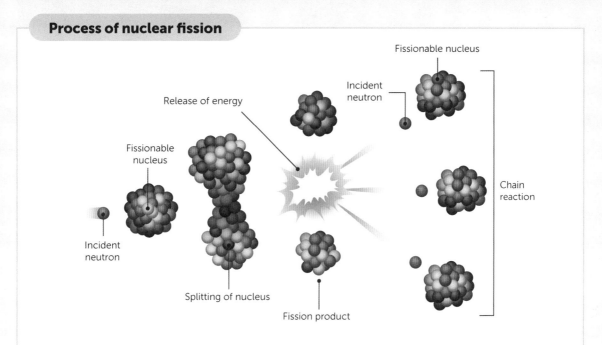

Fissionable nucleus

Incident neutron

Release of energy

Fissionable nucleus

Chain reaction

Incident neutron

Splitting of nucleus

Fission product

The nucleus undergoing fission splits into two smaller nuclei of roughly equal size. The nucleus also emits two or three neutrons, and gamma rays. A large amount of energy is released during this reaction. All of the fission products move so they have kinetic energy.

The released neutrons may go on to split other large nuclei which release more neutrons, starting a **chain reaction**.

In a **nuclear reactor**, the chain reaction from the nuclear fuel is very carefully controlled using a material to absorb some of the neutrons. The energy released is then used for generating electricity.

An uncontrolled chain reaction is extremely dangerous and is the cause of the explosion in a nuclear weapon.

1. What is meant by a chain reaction? [1]
2. Give the term used for the nuclei produced from the fission of an unstable nuclei. [1]
3. Explain what type of chain reaction the diagram above shows. [2]

1. *When products from one reaction move on to start another reaction and this continues.[1]*
2. *Daughter nuclei.[1]*
3. *It shows an uncontrolled chain reaction[1] because three neutrons go on to cause three more nuclei to undergo nuclear fission[1].*

# NUCLEAR FUSION

**Nuclear fusion** happens when two light nuclei join together to form one heavier nucleus.

## Fusion reaction

Deuterium and tritium are isotopes of hydrogen. When these two smaller nuclei undergo nuclear fusion, they join to form one larger nuclei of helium.

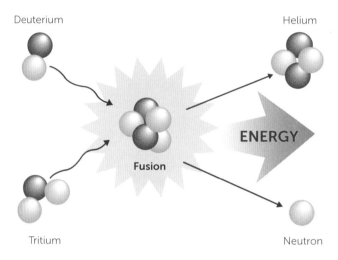

Nuclear fusion reactions happen in stars (including the Sun) under high temperatures and pressure. A huge amount of energy is released. During nuclear fusion, some of the mass may be converted into energy.

1. Describe the difference between nuclear fission and nuclear fusion. [2]

2. Nuclei need to get very close to each other before fusion can happen. Explain why very high temperatures and pressures are needed for nuclear fusion to happen. [3]

1. *The nuclei are positively charged[1] and so repel each other[1]. At high temperatures and pressures, the nuclei will have enough energy to overcome the force of repulsion.[1]*

2. *Both nuclei have a positive charge[1] so as they come to collide, their positive charges will repel each other [1] so they must fuse quickly to overcome this repulsion[1].*

# EXAMINATION PRACTICE

01  One measure of the atomic radius of hydrogen is $4.0 \times 10^{-11}$ m.

01.1 What is the atomic radius of hydrogen in nm? Tick **one** box. [1]

☐ 0.004 nm ☐ 0.04 nm ☐ 4.0 nm ☐ 40 nm

01.2 Calculate the diameter of a hydrogen atom in metres. [1]

02  This represents an atom of fluorine.

$$^{19}_{9}F$$

02.1  Describe this atom of fluorine in terms of subatomic particles. [3]

02.2  Give the atomic number and mass number of an atom that contains 13 protons and 14 neutrons. [2]

Atomic number _____ Mass number _____

02.3  Describe the difference in atomic particles between isotopes of the same element. [1]

03  This question is about the development of the atomic model.

03.1  Describe the plum pudding model of the atom. [2]

03.2  Draw lines to match the evidence from the alpha scattering experiment with the conclusion made about an atom. [2]

| Evidence |
| --- |
| Some alpha particles were repelled. |
| Some alpha particles were reflected back. |
| Most alpha particles passed straight through. |

| Conclusion |
| --- |
| An atom is mainly empty space. |
| The nucleus is positively charged. |
| The atomic mass is concentrated in the centre. |

04  Radioactive isotopes have different half-lives.

04.1  The count rate of a radioactive substance falls from 240 Bq to 60 Bq after 14 years. Calculate the half-life of this substance. [2]

04.2  Another isotope has a half-life of 30 minutes. Calculate the net decline as a ratio after one hour. [2]

04.3  An object is contaminated with radioactive atoms. Explain why the half-life of the radioactive atoms affects the level of hazard. [2]

05 This question is about nuclear radiation.

05.1 Write the **three** main types of nuclear radiation in order from highest to lowest for their:

Range in air: _____ _____ _____ [1]

Ionising power: _____ _____ _____ [1]

05.2 Smoke detectors use a source of nuclear radiation to ionise air in the smoke detector. This allows a small current to flow between electrodes. When smoke gets into the gap, the current reduces, and an alarm sounds.

| Radioactive isotope | Radiation emitted | Half-life |
|---|---|---|
| Americium-241 | Alpha | 432 years |
| Cobalt-60 | Gamma | 5.3 years |
| Thorium-228 | Alpha | 1.9 years |
| Nickel-63 | Beta | 100 years |

The table shows four sources of nuclear radiation.

Evaluate the suitability of each radioactive isotope for use as a source of radiation in a smoke detector. Include descriptions of the properties in your evaluation. [6]

05.3 Complete the nuclear equation for the alpha decay of Americium-241 [3]

$$^{241}_{95}\text{Am} \longrightarrow \underline{\hspace{1cm}} \text{Np} + \underline{\hspace{1cm}}$$

05.4 The atomic number of nickel is 28.
Write a balanced nuclear equation for the beta decay of nickel-63 to copper. [3]

**Physics only**

06 This question is about nuclear radiation.

06.1 Define background radiation. [1]

06.2 Give **one** natural and **one** man-made source of background radiation. [2]

06.3 Explain why the half-life of a radioactive medical tracer is an important consideration in its selection. [1]

06.4 Explain how nuclear radiation can both cause cancer and kill cancer cells. [2]

07 This question is about nuclear fission and nuclear fusion.

07.1 Draw a simple diagram to represent the nuclear fission reaction of a uranium nucleus. Label the particles. [3]

07.2 Define nuclear fusion. [1]

# TOPICS FOR PAPER 2

## Information about Paper 2:

### Separate Physics 8461:

**Written exam: 1 hour 45 minutes**
**Foundation and Higher Tier**
**100 marks**
**50% of the qualification grade**
**All questions are mandatory**

### Trilogy 8464:

**Written exam: 1 hour 15 minutes**
**Foundation and Higher Tier**
**70 marks**
**16.7% of the qualification grade**
**All questions are mandatory**

**Specification coverage**

The content for this assessment will be drawn from Topics 5–8. Forces; Waves; Magnetism and electromagnetism; and Space physics.

**Questions**

A mix of calculations, multiple-choice, closed short answer and open response questions assessing knowledge, understanding and skills.

Questions assess skills, knowledge and understanding of Physics.

# SCALAR AND VECTOR QUANTITIES

A **quantity** is an amount or a measure of something. In Physics, these are things that can be measured.

## Quantities

There are two types of quantities:

- **Scalar** quantities – which only have a value for size, called a **magnitude**, and
- **Vector** quantities – which have a **magnitude** and also a **direction**.

Vector quantities can be represented by arrows.

- The **length** of the arrow represents the **magnitude** of the vector quantity.
- The **direction** of the arrow represents the **direction** of the vector quantity.

## Examples of scalar and vector quantities

| Scalar qualities | Vector qualities |
|---|---|
| Mass | Force |
| Speed | Velocity |
| Distance | Displacement |
| Energy | Acceleration |
| Temperature | Momentum |

Two runners are running at the same speed, but in different directions. Their speeds are the same but their velocities are different. One runner has a velocity of 5 m/s to the left and the other runner has a velocity of 5 m/s to the right.

If we define moving to the right as being in the positive direction, the runner on the left has a velocity of −5 m/s.

1. Explain, giving examples, the difference between speed and velocity. [2]

2. Explain why weight is a vector quantity. [2]

3. An object has a velocity of 4 m/s. What is the velocity of an object moving at the same speed in the opposite direction? [1]

1. *Speed only has a magnitude, for example 10 m/s[1], velocity also has a direction. for example 10 m/s south[1].*

2. *Weight is a force[1] which has both magnitude and direction.[1]*

3. *−4 m/s[1]*

5 m/s          5 m/s

# CONTACT AND NON-CONTACT FORCES

Forces are pushes and pulls that act on an object due to an interaction with another object.

## Types of force

Force is a vector quantity, so it has size and direction. Forces are represented by arrows, often drawn to scale. The longer the arrow, the larger the force.

The size and direction of a force is determined by the force that is acting and how it is acting.

There are two types of forces:

- **Contact forces** – when objects are touching.
- **Non-contact forces** – when objects are not touching or are separated.

## Non-contact forces

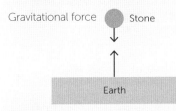

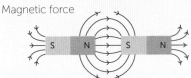

## Contact forces

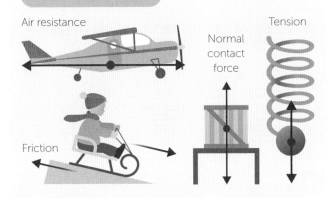

3. Explain the difference between a gravitational force and a magnetic force. [2]

   *3. A gravitational force always attracts[1] but a magnetic force can attract or repel.[1]*

## Interaction pairs

A book is sitting on a table. The weight of the book is balanced by the reaction force from the table. This force is at right angles to the surface and is called the **normal contact force** from the table. There is an interaction between the two objects which produces a force on each object. 'Normal' means at 90°.

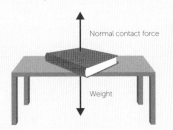

1. Explain why an electrostatic force is a non-contact force. [2]
2. Identify the other force in each of the interaction pairs shown in the contact force diagrams. [4]

   *1. There will be a force between two charged objects when they are brought close together but not touching.[1] The objects do not have to be touching for the force to be exerted.[1]*

   *2. Tension: weight[1]; Air resistance: pushing force or thrust[1]; Friction: pushing force[1]; Normal contact force: weight[1].*

# GRAVITY

## Mass and weight

**Mass** is the amount of matter that an object contains. Mass is constant regardless of the strength of the gravitational field it is in. It is a scalar quantity.

**Weight** is the **force** acting on a mass due to gravity. Weight changes and depends on the strength of the gravitational field the mass is in.

The weight and mass of an object are directly proportional to each other. The weight of an object can be calculated using the equation:

### weight = mass × gravitational field strength

$W = m\,g$

$W$ = weight in newtons, N

$m$ = mass in kilograms, kg

$g$ = gravitational field strength in newtons per kilogram, N/kg

| $W$ | | |
|---|---|---|
| $m$ | × | $g$ |

You need to be able to recall and apply this equation.

## Force of gravity

The force of gravity near or on the Earth is due to the gravitational field around the Earth.

Forces, including weight, can be measured using a **newtonmeter**, which is a calibrated spring-balance. The spring stretches a fixed distance for each newton.

The weight of an object is considered to act at a single point known as the **centre of mass** of the object. This is useful when drawing diagrams representing forces.

$g$ on the surface of the Earth will be given in an exam as 9.8 N/kg, or rounded to 10.0 N/kg. An object is sometimes called a body.

1. Calculate the weight of a 0.5 kg mass on the moon.
   Use gMoon = 1.6 N/kg.    [2]
2. An object exerts a force of 6.9 N on Earth. Calculate the mass of the object in g. Use $g$ = 9.8 N/kg.    [4]

> 1.  $W = m \times g = 0.5 \times 1.6^{[1]} = 0.8\ N^{[1]}$
> 2.  $W = m \times g$
>     $6.9 = m \times 9.8^{[1]}$
>     $m = 6.9 / 9.8^{[1]} = 0.7\ kg^{[1]} = 700\ g^{[1]}$

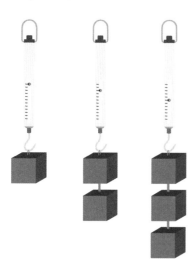

# RESULTANT FORCES

There is usually more than one force acting on an object. The different forces can be replaced by a single force that has the same effect as all of the other forces acting together. This single force is called the **resultant force**.

## Predicting behaviour

Finding the resultant force simplifies diagrams and calculations. This helps predictions to be made about how an object might behave.

- When a **pair of different sized forces** are in opposite directions, an **unbalanced force** causes a **resultant force**.
- When a **pair of equal sized forces** are in **opposite directions**, this is a **balanced force** and there is no **resultant force**.

## Calculating the resultant of parallel forces acting in a straight line

2 kg

$F_1 = 6N$

$F_2 = 4N$

2 kg

$F_3 = 3N$

$F_1 = 6N$

As the forces are in the same direction they can be added together.

$F_1 + F_2 = 6 + 4 = 10$ N

$F_{right} = 10$ N

$F_3$ is acting to the left and is in the opposite direction to $F_1$.

Assume that the positive direction is to the right. So, $F_1 = -3$ N

Add the forces together to find the resultant force.
$F_{resultant} = F_1 + F_3 = 6$ N + −3 N = 3 N (to the right).

Remember to give the direction of the force. Forces in opposite directions have opposite values.

The diagram shows three blocks in water.

(a) Calculate the resultant force on each block.          [3]

(b) Describe the movement of the blocks.          [1]

   (a) Block A: resultant force = 3 + −3 = 0 N[1]
     Block B: resultant force = 4.5 + −3 = 1.5 N upwards.[1]
     Block C = resultant force = 3 + −4.5 = −1.5N, so 1.5N downwards.[1]
   (b) Block A: stationary,  Block B: moves upwards,
     Block C: moves downwards.[1]

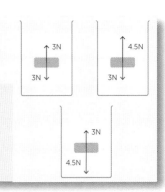

# RESOLVING FORCES

**Vector diagrams** and **scale drawings** can be used to solve problems when forces act at angles to each other.

## Resolving forces

The arrows are best drawn on graph paper using a ruler, protractor and sharp pencil to show both the magnitude and the direction of the forces.

> Remember, force is a vector quantity so has size and direction. Forces are represented by arrows.

Choose a suitable scale for the magnitude of each force and add an arrow to show the direction it is in. Any angle is given as a value 'to the horizontal' or 'to the vertical.'

## Resolving a single resultant force

A single force acting at an angle can be resolved into two components acting at right angles to each other: a **horizontal force** and a **vertical force**. The two forces have the same effect as the single force.

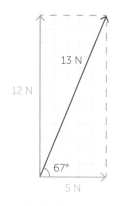

To resolve a force into horizontal and vertical components draw a scale diagram on squared paper. Use a protractor to draw the angle and measure its length using a ruler. Then measure the lengths of the components. The diagram shows a force of 13 N at an angle of 67° to the horizontal. The force has been drawn on squared paper with one large square representing 5 N. Count the number of squares to find the horizontal and vertical components: $F_h = 5$ N and $F_v = 12$ N.

This vector diagram shows how the force in red can be resolved into two forces at right angles to each other. Force $F$ is acting at an angle of $\theta°$ to the horizontal. It can be resolved into a vertical component, $F_y$ and a horizontal component $F_x$.

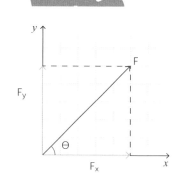

## Resolving two perpendicular forces into single resultant force

You can resolve a horizontal and vertical force into a single resultant force by drawing a scale diagram. Draw the horizontal force first. Then draw the vertical force, starting it at the end of the horizontal force. Measure the length of this arrow using a ruler. Measure the size of the angle using a protractor. The scale diagram shows a horizontal force of 6 N and a vertical force of 4 N. The resultant force has a magnitude of 7.2 N acting at 34° to the horizontal.

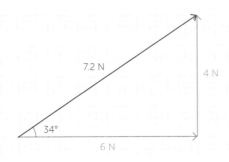

1. Draw a vector diagram to resolve a force of 58 N at 31° to the horizontal into a horizontal force and a vertical force. [2]

2. Use a scale drawing to find the resultant of a force of 20 N acting to the right and a force of 40 N acting upwards. [2]

1. *Scale drawing shows:*
   *horizontal component: 50 N to the right[1];*
   *vertical component: 30 N upwards[1].*

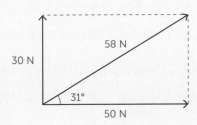

2. *Resultant: 45 N[1] at 63° to the horizontal[1].*

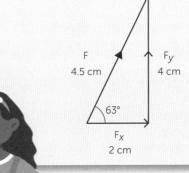

# FREE BODY DIAGRAMS

A **free body force diagram** is a simplified diagram showing the forces acting on an object shown as a simple box or a dot. The force arrows act away from the centre of the box or dot.

## Balanced forces

The air resistance is equal and opposite to the weight of the parachutist so there is no net force on the parachutist and the forces are balanced.

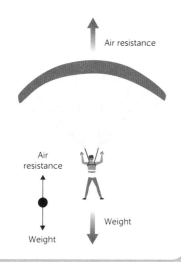

## Unbalanced forces

The weight of the aeroplane is equal and opposite to the lift of the plane. So there is no net vertical force.

The thrust from the engines of the aeroplane is greater than the air resistance. So the horizontal forces are unbalanced and there is a resultant force to the left.

The same force can sometimes be described using different words. For example, **drag** and **air resistance**; **forward force** and **thrust**. The situation of the object, and the direction of the force arrow, will make what the force represents clearer.

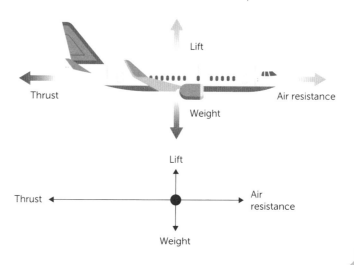

Figure X, below, shows a submarine moving forward at a constant depth. Draw a free body diagram of this submarine. [2]

[1] Horizontal
[1] Vertical

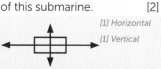

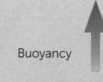

Buoyancy

Water resistance

Engine force

X

Weight

# WORK DONE AND ENERGY TRANSFER

**Mechanical work** is done on an object when a force moves the object through displacement. The amount of work done depends on the force used and the distance the object is moved.

## Calculating work done

The work done by a force on an object can be calculated using the equation:

$$\text{work done} = \text{force} \times \text{distance}$$

$W = F\,s$

$W$ = work done in joules, J

$F$ = force in newtons, N

$s$ = distance moved in metres, m

This distance is the **distance moved along the line of action of the force**.

When a force of 1 N causes a displacement of 1 m, 1 J of work is done or 1 J of energy is transferred.

> You need to be able to recall and apply this equation.

**1 joule (J) = 1 newton-metre (Nm)**

## Examples of energy transfers when work is done

Here are some energy transfers that happen when work is done:

- A person lifting a mass against gravity transfers energy from their own chemical store to the gravitational potential energy store of the mass.
- Braking in a car transfers energy from the kinetic store of the car to the thermal energy store of the brakes.

Work done against the frictional forces acting on an object causes a rise in the temperature of the object.

> Electrical work also transfers energy and is covered on **page 31**.

---

1. A person with a weight of 600 N goes up a vertical distance of 40 m on a hill. Calculate the work done by the person in kJ. [3]

2. Explain why brake pads get warm when used to stop a bicycle. [3]

   1. *W = F s*
      *= 600 × 40[1] = 24 000 J[1] = 24 kJ[1]*

   2. *Work is done by the friction force between the brakes and the wheel rims or brake discs.[1] As the bike slows energy is transferred by friction from the kinetic energy store of the bike[1] to the thermal energy store of the brake pads (so their temperature increases)[1].*

# FORCES AND ELASTICITY

**Objects** can change shape when forces are applied.

## Deformation

Forces can change the shape of a stationary object by **stretching**, **bending** or **compressing** the object. The change of shape is called **deformation**. There are two types of deformation.

- **Elastic** deformation is when the object goes back to its original shape when the force is removed.

- **Inelastic** deformation means that the material does not return to its original shape when the force is removed. There is a permanent change of shape.

To stretch, bend or compress an object, two forces need to be applied to the object. If just one force is applied, the object would just move and would not change shape.

Stretching needs two forces pulling away from each other. Compression needs two forces pushing towards each other. Bending needs one force acting clockwise and the other anticlockwise.

The **extension** of an elastic object (such as a spring) is **directly proportional** to the force applied, provided that the **limit of proportionality** is not exceeded. This means it has not been inelastically deformed. This equation relates the force to the resulting extension (or compression) of an elastic object:

$$\text{force} = \text{spring constant} \times \text{extension}$$

$F = k\,e$

$F$ = force in newtons, N

$k$ = spring constant in newtons per metre, N/m

$e$ = extension in metres, m

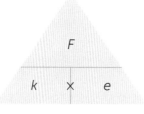

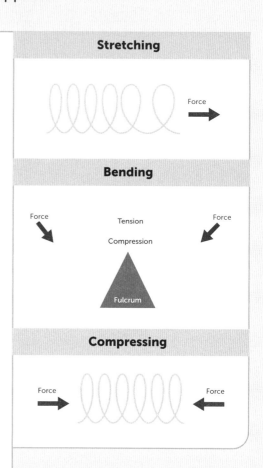

**Stretching**

Force

**Bending**

Force         Tension         Force

Compression

Fulcrum

**Compressing**

Force                                    Force

You need to be able to recall and apply this equation.

## Energy stored in a spring

When an elastic object is stretched or compressed by a force, work is done. This means that an energy transfer takes place and energy is transferred to the elastic potential energy store of the object. As long as the deformation is elastic, the work done on a spring is equal to the elastic potential energy stored.

The work done in stretching (or compressing) a spring (up to the limit of proportionality) can be calculated using the equation:

**elastic potential energy = 0.5 × spring constant × (extension)²**

$$E_e = \frac{1}{2} k e^2$$

> You need to be able to select this equation from the equation sheet and apply it.

1. Complete the diagram to show the force needed to stretch the rubber. [2]

2. A force of 1.8 N is applied to a spring with a spring constant of 12 N/m. The spring goes back to its original shape when the force is removed. Calculate the extension of the spring. [3]

3. The original length of a spring is 10 cm. A force of 20 N is applied to the spring which causes its length to increase to 15 cm. Calculate the spring constant of the spring. [3]

4. A spring is extended by 0.1 m. The spring has a spring constant of 30 N/m. Calculate the work done to stretch the spring. [2]

1. *1 m arrow in opposite directions[1], of the same size[1].*

2. *F = k e     1.8 = 12 × e[1]*
   *e = 1.8 / 12[1] = 0.15 m[1]*

3. *Extension = extended length − original length = 15 − 10 = 5 cm = 0.05 m[1]*
   *F = k e     20 = k × 0.05[1]*
   *k = 20 ÷ 0.05 = 400 N/m[1]*

4. *Work done = $E_e = \frac{1}{2} k e^2$*
   *= 0.5 × 30 × 0.1²[1]²[1] = 0.15 J[1]*

## Linear and non-linear relationship between force and extension

Overstretched spring

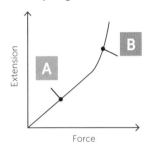

**A**   Elastic deformation: Straight line so it is a linear relationship.

    The line goes through the origin (0,0) so extension is directly proportional to force.

**B**   Inelastic deformation: Limit of proportionality is exceeded.

Elastic band

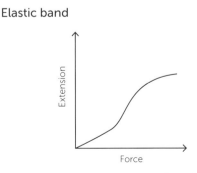

Curved line so a non-linear relationship.

# REQUIRED PRACTICAL 6 (18)

## Force and extension of a spring

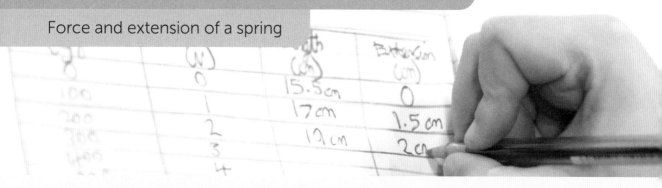

This practical activity helps you to calculate a spring constant and use a graph to find the weight of an unknown object.

> Remember to convert the extension to metres if you measured it in centimetres.

## Experiment method

This apparatus is used to measure the extension of a spring as the force is increased by adding masses one at a time. Each force is calculated using $F = m\,g$.

Make sure the readings on the metre rule are taken horizontally and not at an angle.

Record the original length of the spring when $F = 0$ N. It does not matter if the end of the spring does not line up with zero on the ruler when $F = 0$ N. Take two readings, one at each end of the spring and subtract one from the other to find the original length.

The extension is the total increase in length from the original length of the spring each time the force is increased.

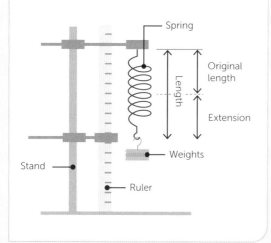

## Analysis of results

Plot a line graph of **extension against force**. This means 'Extension of spring in m' is on the y-axis and 'Weight in N' is on the x-axis.

It is a straight-line graph so its equation will be in the form $y = mx + c$.

where m = gradient and $c$ = y-intercept
$c = 0$ as the line goes through the origin.
$y = e$ and $x = F$ so $y = mx$ can be written as
$e = mF$
gradient, $m = e / F$
Using $F = ke$, gradient = e / F = 1 / k

So, the spring constant can be calculated by determining the gradient, and finding the reciprocal of the gradient..

### spring constant = 1 / gradient

The graph shows that the extension of a spring is directly proportional to the force applied. This is only for a linear part of the graph before the limit of proportionality has been exceeded. The spring undergoes elastic deformation.

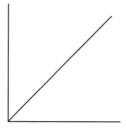

### Calculating the spring constant from a graph

Draw a triangle on the line of best fit to determine the gradient.

If you have plotted the extension in centimetres, you will need to convert it to metres when finding the gradient.

Gradient = (0.06 – 0.02) / (4.8 – 1.6)
 = 0.04 / 3.2 = 0.0125

Spring constant = 1 / gradient
 = 1 / 0.0125
 = 80 N/m

You can also use the graph to calculate the elastic potential energy stored in the spring. It is the area under the graph.

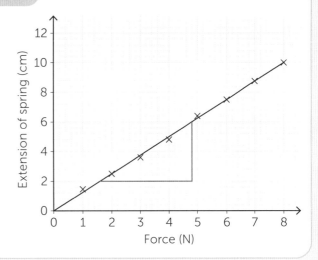

### Using the graph to determine the weight of an unknown object

Hang the object on the spring and measure the extension. Use the graph to determine the force at that extension, this will be the weight of the object. It could be converted to mass by dividing it by g, the gravitational field strength.

1. Explain which spring, Spring A or Spring B, is the stiffest. Use the graph in the figure below to help.  [3]

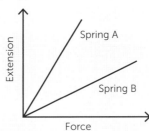

2. Determine the weight of an object which extends the spring in the diagram by 9 cm. Show your working on the graph at the top of the page.  [2]

   1. *Spring B is the stiffest spring.[1] A stiffer spring has a higher spring constant, k.[1] The spring constant is inversely proportional to the gradient[1] so the higher the spring constant, the lower the gradient[1].*

   2. *Horizontal line drawn from 9 cm extension.[1] Weight = 7.2 N.[1]*

# MOMENTS, LEVERS AND GEARS

Forces can cause an object to turn or rotate about a **pivot** in a clockwise or anticlockwise direction. The **turning effect** of a force is called the **moment** of the force.

## Moment

The moment of a force can be calculated using the equation:

moment of a force = force × distance

$$\frac{M}{F \times d}$$

$M = F \times d$

$M$ = moment in newton-metres, Nm

$F$ = force in newtons, N

$d$ = the **perpendicular distance from the pivot to the line of action of the force**.

Perpendicular means the distance is at right angles (90°) to the force

### Example

$d$ = perpendicular distance from the pivot to the line of action of the force in metres, m

The longer the spanner handle, the less force is needed to produce the same turning effect.

Opening a door closer to the hinge is more difficult than opening it further away from the hinge as more force is needed to produce the same turning effect.

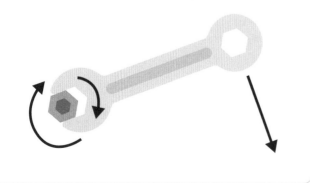

You need to be able to recall and apply this equation.

## Rotational effects

When objects are **balanced** about a pivot:

The **total clockwise moment** ↻ equals the **total anticlockwise moment** ↺. When the anticlockwise moment is larger, the left-hand side of a seesaw goes down. When the clockwise moment is larger, the right-hand side goes down.

The **rotational effects** of forces can be **transmitted** using simple systems like **levers** and **gears**.

When the plank is balanced about the pivot:

$F_1 \times d_1 = F_2 \times d_2$

So, an unknown force or distance can be calculated if the other three values are known.

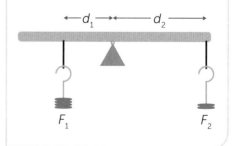

## Simple gear system

Gears are toothed wheels (cogs) that work together to transmit rotation force and motion.

Each gear in a series reverses the direction of motion.

When the small cog moves clockwise, the middle cog moves anticlockwise. The middle cog will rotate more slowly than the smaller cog because its circumference is larger. Because it is rotating more slowly, it transmits a higher force.

The same principle is used in bicycle gears. When you go up hill, you change into a lower gear which means that the gear on the wheel is larger than the wheel on the pedals. The cyclist can then transmit more force to the wheels to get up the hill.

## Simple lever system

A lever uses the moment of a force. The man is applying an anticlockwise force (effort) at a distance from the pivot, giving an anticlockwise moment. This moment is greater than the clockwise moment from the load, so the load moves. The load is much bigger than the force applied by the man, but it only moves a small distance. The lever acts as a force multiplier.

Other examples of a lever are a wheelbarrow and scissors.

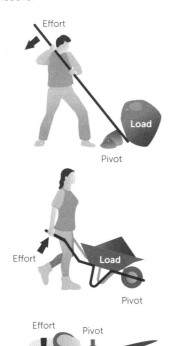

1. Explain why scissors transfer a bigger force when you cut something near the pivot rather than at the tips of the scissors. [2]

2. A force of 45 N is applied at a perpendicular distance of 0.6 m from a pivot. Calculate the moment of the force. Include the unit. [3]

3. A child weighing 240 N sits 1.5 m from the pivot on one side of a seesaw. Another child with a weight of 300 N sits on the opposite side of the seesaw. Calculate the distance this child must sit from the pivot to balance the seesaw. [3]

4. The large cog in the diagram above is turned clockwise.
Describe the movement of the small cog. [2]

1. As the moment is equal to the force multiplied by the perpendicular distance from the pivot[1], when the distance is shorter, the force will be higher[1].

2. $M = F \times d = 45 \times 0.6$[1] $= 27$[1] $Nm$[1]

3. $F_1 \times d_1 = F_2 \times d_2$
$240 \times 1.5 = 300 \times d$[1]
$d = 360 / 300$[1] $= 1.2\ m$[1]

4. The middle cog moves anticlockwise, so the small cog moves clockwise.[1] The small cog will rotate much faster and with less force.[1]

# PRESSURE IN A FLUID 1

**Pressure** is the force per unit area. The pressure in a fluid causes a **normal force** (a force at right angles) to act at any surface it touches. A fluid can be either a liquid or a gas.

## Calculating pressure

The pressure at the surface of a fluid can be calculated using the equation:

$$\text{pressure} = \frac{\text{Force normal to a surface}}{\text{Area of that surface}}$$

$p = \dfrac{F}{A}$

$p$ = pressure in pascals, Pa

$F$ = force in newtons, N

$A$ = area in metres squared, m²

You need to be able to recall and apply this equation.

## Pressure in fluids

A pressure in a fluid acts in all directions. This means that pressure is transmitted throughout the fluid. In a piston system, where a liquid is used, this principle can be used to lift large objects, such as a hydraulic car jack. A force applied to the smaller piston allows a larger force to be exerted in the larger piston.

| Air pressure in a balloon | Pressure in a fluid | Pistons |
|---|---|---|
| | | |

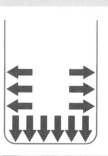

1. Calculate the downward force, *F*, needed to act on the small piston to balance the car on the large piston.    [4]

Pressure exerted by the car = $\frac{F}{A}$

= 2000 / 2[1] = 10 000 Pa[1]

So force needed on smaller piston:

10 000 = F / 0.04[1]

F = 400 N[1]

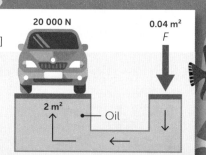

# PRESSURE IN A FLUID 2

The pressure in a column of liquid is dependent on the **depth** in the liquid. This is because the further down in the liquid, the greater the volume and weight of the liquid above.

## Pressure at different depths

The pressure in a liquid increases with depth. The pressure at the bottom hole in the bottle is higher than at the middle hole, which is higher than at the top hole. So the water at the bottom hole will shoot out further.

The pressure at a certain depth of a liquid with a higher **density** is higher than in liquid with a lower density at the same depth. The mass of the liquid is higher because its density is higher, so the weight of the liquid and the pressure in the liquid is higher. See **4.3.1.1** on **page 37**.

High pressure

Medium pressure

Low pressure

2. The diagram shows two identical containers with a hole at the same height.
   Explain what the diagram shows about the densities of cooking oil and water. [2]

Cooking oil    Water

3. The depth of water behind a dam is 15 m. Calculate the difference in pressure between the bottom of the dam and 2 m below the surface of the water. The density of water is 1000 kg/m³. Use g = 9.8 N/kg. [4]

## Calculating pressure

The pressure due to a column of liquid can be calculated using the equation:

**pressure = height of the column × density of the liquid × gravitational field strength**

$p = h \rho g$

$p$ = pressure in pascals, Pa

$h$ = height of the column in metres, m

$\rho$ = density in kilograms per metre cubed, kg/m³

$g$ = gravitational field strength, g, in newtons per kilogram, N/kg

> You need to be able to recall and apply this equation.

2. *For the same height, the water shoots out further than the oil, so the pressure in the water is greater than in the oil[1] and so water has a greater density than cooking oil.[1]*

3. *$p = h \rho g$*
   *At 15 m, p = 15 × 1000 × 9.8[1] = 147 000 Pa[1]*
   *At 2 m, p = 2 × 1000 × 9.8 = 19 600 Pa[1]*
   *Difference in pressure =*
   *147 000 − 19 600 = 127 400 Pa[1]*

# FLOATING AND SINKING

An object fully or partially immersed in a fluid will experience a greater pressure on its bottom surface than on its top surface because pressure increases with depth in the fluid.

This means there is a resultant force created upwards on the object called **upthrust**.

## Upthrust

The relationship between the upthrust and the weight of the object determines whether the object **floats** or **sinks**.

An object will:

- **float** if the upthrust is greater than, or equal to, its weight.
- **sink** if the upthrust is less than its weight.

The upthrust depends on the densities of both the liquid and the object.

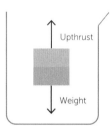

Greater density means greater weight for the same volume.

1. Explain why ice floats in water.  [3]
2. Explain why the aluminium is sitting in the bottom of the container.  [2]

   *1. The density of ice is lower than the density of water[1] so the ice can displace its own weight in water[1] and floats because the upthrust equals the weight of the ice[1].*

   *2. It has a higher density than the water[1] so it sinks[1].*

## Floating and sinking

Assume an object is placed in water. The water rises up the container **displacing** a volume of water equal to the volume of the object in the water.

The **weight of the displaced water** is equal to the size of the **upthrust** acting on the object.

If the object has a **higher density** than water, then the volume of water it displaces will have a lower weight than the object. Hence the upthrust is less than the weight of the object and it **sinks** e.g. aluminium.

If the object has a **lower density** than water, then the volume of water it displaces will have a greater weight than the object. Hence the upthrust is greater than the weight of the object and it **floats** e.g. cork and wood.

The cork floats higher in the water than wood because cork has a lower density than wood.

Cork          Wood          Aluminium

# ATMOSPHERIC PRESSURE

The **atmosphere** is a thin layer (compared to the size of the Earth) of air around the Earth. It gets less dense with increasing altitude.

## Simple model of the Earth's atmosphere

**Atmospheric pressure** is caused by air molecules colliding with a surface. The surface could be on the Earth, or an object in the atmosphere such as an aeroplane.

As the height of a surface above ground level increases:

- the number of air molecules above the surface decreases
- the weight of air above the surface decreases.

There is always less air above a high surface than there is at a lower height. So atmospheric pressure decreases as height increases.

You can link this to the equation for calculating the pressure at a depth in a fluid.

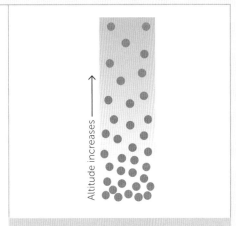

**Altitude** is the height above sea level on Earth.

## Simple model of atmospheric pressure

The molecules in a gas move continuously at a high speed. The molecules have many collisions. Sometimes they collide with each other and sometimes they collide with a surface. It is the collisions of the molecules with a surface that creates the atmospheric pressure.

The atmosphere above a surface can be thought of as a column of air. The column exerts a pressure due to the weight of the air which acts at 90° to the surface area. Higher up the column, the density of air is less, so there are fewer molecules of gas in the same volume. There are fewer collisions per second with a surface and so less pressure on the surface.

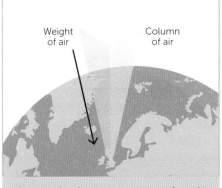

Atmospheric pressure at surface is about 100 000 Pa, or 1 atmosphere

1. Explain why the atmospheric pressure at the top of Mount Everest is much lower than it is at sea level.                                                                                        [2]

2. Describe the relationship between altitude and atmospheric pressure.                    [1]

   1. *At the top of Mount Everest, the column of air above you is much smaller, so the density of the atmosphere is much smaller.[1] There are fewer collisions so the pressure is lower.[1]*

   2. *The higher the altitude, the lower the atmospheric pressure.[1]*

# DISTANCE AND DISPLACEMENT

**Distance** and **displacement** are both ways of describing how far an object has moved from its starting point.

## Properties of distance and displacement

| Distance | Displacement |
|---|---|
| The total length of the path travelled by a moving object from start to finish | The shortest distance between the start and finish points of a moving object |
| Any shape line | Straight line |
| Scalar quantity | Vector quantity |
| No direction | The direction of the straight line is described in terms of an angle or a point of the compass |
| Measured in metres, m | Measured in metres, m |
| Will always have a positive value | Values can be positive or negative depending on the direction moved |
| Will still be total distance moved when an object has moved from one point and returned to the same place | Will be zero when an object starts and finishes at the same place |

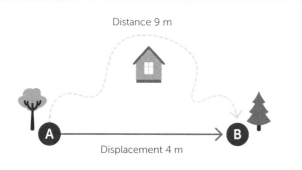

Distance 9 m

Displacement 4 m

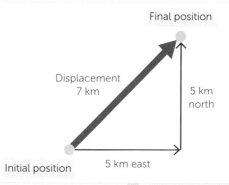

Final position

Displacement 7 km

5 km north

Initial position

5 km east

1. Describe the difference between displacement and distance. [1]

2. Describe the displacement of the object above right. [1]

1. Distance only has magnitude, displacement has a magnitude and a direction.[1]

2. The displacement is 7 km north east.[1]

# SPEED

Speed is a **scalar** quantity so has size but no direction. Moving objects usually travel with **non-uniform motion** so speed is usually given as a **mean** value over a certain distance or time.

## Typical speeds

The speed a person walks, runs or cycles depends on many factors such as the distance travelled, the person's age and fitness, the terrain and the weather.

Vehicle speeds in the UK are usually given in miles per hour (mph). However, in physics, speed is measured in metres per second (m/s).

The table shows some typical values for speed. All can vary.

| Example | Typical speed in m/s |
|---------|---------------------|
| Walking | 1.5 |
| Running | 3 |
| Cycling | 6 |
| Car | 13–31 |
| Train | 50 |
| Plane | 250 |
| Sound | 330 m/s |

## Distance travelled, speed and time

For an object moving at constant speed, the distance travelled in a specific time can be calculated using the equation:

You need to be able to recall and apply this equation.

**distance travelled = speed × time**

$s = v\,t$

$s$ = distance in metres, m

$v$ = speed in metres per second, m/s

$t$ = time in seconds, s

Note that $s$ represents distance here – not speed or seconds. $v$ is also used for velocity.

1. A cyclist travelled at a constant speed of 5 m/s for 25 m. Calculate the time taken. [2]
2. A car journey of 42 km takes 50 minutes. Calculate the mean speed of the car. [4]

   1. $s = v \times t$   $25 = 5 \times t$[1]   $t = 25/5$[1] $= 5$ s[1]
   2. 50 minutes = $50 \times 60$ seconds = 3000 s[1]
      $s = v \times t$   $42\,000 = v \times 3000$[1]
      $v = 42\,000 / 3000$[1] $= 14$ m/s[1]

# VELOCITY

Velocity is a vector quantity representing speed in a given direction.

### Example

Both cars are travelling at the same speed of 20 m/s.

The orange car is travelling at a velocity of 20 m/s to the right. The red car is travelling at a velocity of 20 m/s to the left.

We need to define one direction as positive – this is usually taken to be to the right or upwards.

So we can say that the orange car has a velocity of +20 m/s and the red car has a velocity of −20 m/s.

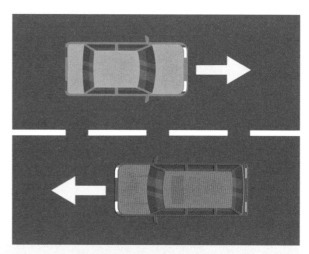

**Direction can be described as: right or left, up or down, north, south, east or west**

### Motion in a circle    Higher Tier

This car is travelling at a constant speed in a circle. The speed is constant because it takes the same time to travel the same distance.

However, the direction of the car changes constantly as it moves in a circle. So the velocity is constantly changing because the direction of the car is constantly changing.

Motion in a circle involves constant speed but changing velocity.

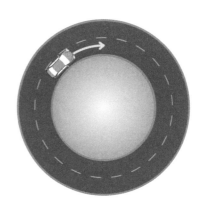

1. Describe the difference between speed and velocity. [1]

2. A student takes 5 minutes to walk 510 m from west to east along a straight beach. Calculate the mean velocity of the student. [4]

1. *Speed only has magnitude and is a scalar, velocity is speed in a given direction and is a vector.[1]*

2. *$t$ = 5 minutes = 5 × 60 = 300 s[1]*
   *$s = v \times t$    510 = $v$ × 300[1]*
   *$v$ = 510/300[1] = 1.7 m/s eastwards.[1]*

# THE DISTANCE-TIME RELATIONSHIP

## Distance-time graphs

A distance-time graph shows how the distance of an object moving in a straight line changes over time. The **gradient** of the graph equals the **speed** of the object.

 speed = gradient = change in distance ÷ change in time

**Uniform motion (top):** The car travels the same distance in each equal time interval so has a constant speed.
The gradient is 40 / 4 = 10 so speed = 10 m/s

**Non uniform motion (bottom):** The car travels a different distance in each equal time interval so the speed changes during the journey shown on this graph.
From A to B the gradient is 20/1 = 20 so the speed = 20 m/s
From B to C the gradient is 10/2 = 5 so the speed = 5 m/s
From C to D the gradient is 10/1 = 10 so the speed = 10 m/s

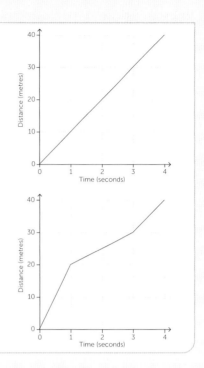

Note that the mean speed for the whole journey is 40 / 4 = 10 m/s as the car has travelled the same total distance in the same time as in the uniform motion graph.

## Distance-time graph for an accelerating object        Higher Tier

When an object is **accelerating** its distance-time graph will be curved. The speed of the object at any particular time can be determined by drawing a **tangent** on the graph and finding its **gradient**.

To work out the speed at 15 seconds, draw a tangent to the graph.

Gradient = change in distance ÷ **change in time**
= (40 − 0) ÷ (23 − 8) = 40 ÷ 15 = 2.7 m/s

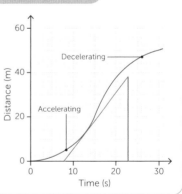

1. Describe what a horizontal line on a distance-time graph represents. [1]
2. A person walks 60 m in 30 s, stops for 5 s, then walks a further 60 m in 20 s. Draw a distance-time graph for this journey. [3]

   1. *The object is stationary as it has travelled no distance during the time.*[1]
   2. *Suitable scale.*[1] *Correctly labelled axes.*[1] *Line drawn from (0,0) to (30,60) to (35,60) to (55,120).*[1]

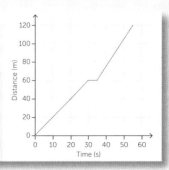

# ACCELERATION

**Acceleration** is a vector quantity representing the rate that an object changes velocity. An object that is slowing down is **decelerating**.

## Calculating acceleration

The average acceleration of an object can be calculated using the equation:

$$\text{acceleration} = \frac{\text{change in velocity}}{\text{time taken}}$$

$$a = \frac{\Delta v}{t}$$

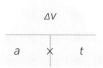

$a$ = acceleration in metres per second squared, m/s²

$\Delta v$ = change in velocity in metres per second, m/s

$t$ = time in seconds, s

An object falling freely under gravity near the Earth's surface has an acceleration of ~9.8 m/s².

> You need to be able to recall and apply this equation.

To **estimate** the magnitude of **everyday accelerations**, make a sensible approximation. For example, you can estimate the acceleration of a typical car when it accelerates from 0 to 100 km/h (about 30 m/s). Make a sensible approximation of the time taken to accelerate, say 10 s. So an estimate of the car's acceleration is (30 − 0)/ 10 = 3 m/s²

So the acceleration of a typical car is ~3 m/s²

> ~ means approximately equal to

## Equation for uniform acceleration

This equation links uniform acceleration, distance and velocity:

$$(\text{final velocity})^2 - (\text{initial velocity})^2 = 2 \times \text{acceleration} \times \text{distance}$$

$$v^2 - u^2 = 2\,a\,s$$

$v$ = final velocity in metres per second, m/s

$u$ = initial velocity in metres per second, m/s

$a$ = acceleration in metres per second squared, m/s²

$s$ = distance in metres, m

1. A cyclist increases velocity from 5.0 m/s to 11 m/s with an acceleration of 1.8 m/s². Calculate the time taken for this acceleration. [3]

2. A sprinter accelerates from stationary with an acceleration of 7.5 m/s² for 5.0 m. Calculate the velocity reached by the athlete. [3]

1. $a = \Delta v/t$
   $1.8 = (11 - 5)/ t$ [1]
   $t = 6/1.8$ [1] $= 3.3$ s [1]

2. initial velocity = 0 m/s
   $v^2 - u^2 = 2\,a\,s$
   $v^2 - 0 = 2 \times 7.5 \times 5$ [1]
   $v = \sqrt{75}$ [1] $= 8.7$ m/s. [1]

# TERMINAL VELOCITY

An object falling through a fluid initially accelerates due to the force of gravity. As the velocity increases, the drag force increases. Eventually the resultant force will be zero and the object moves at a constant speed in one direction. This constant speed is its **terminal velocity** and is the maximum speed it can reach under those conditions.

**Example**    **Physics only**

Think about a skydiver jumping out of an aeroplane. At the moment the skydiver jumps, the air resistance, or drag, is zero; there is a large resultant force downwards and acceleration is at a maximum.

As the velocity of the skydiver increases, the air resistance increases, and the resultant force decreases. The acceleration decreases.

The air resistance increases until the resultant force on the skydiver is zero, and so the acceleration is also zero. The skydiver cannot travel any faster and is travelling at terminal velocity.

> **Terminal velocity**
> Drag force = Weight
> $F_d = F_g$

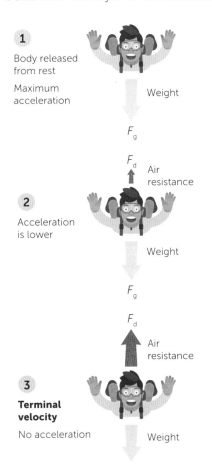

**1**
Body released from rest

Maximum acceleration

Weight

$F_g$

$F_d$    Air resistance

**2**
Acceleration is lower

Weight

$F_g$

$F_d$

Air resistance

**3**
**Terminal velocity**

No acceleration

Weight

$F_g$

## Velocity-time graph

This is a velocity-time graph for an object falling through a liquid that reaches terminal velocity.

**Straight line upwards** – object accelerates and speeds up

**Curved section** – acceleration decreases, object speeding up at a lower rate

**Horizontal** – terminal velocity reached, acceleration is zero

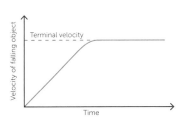

1. Describe speed and acceleration at terminal velocity.   [2]
2. **Physics only:** Explain why a raindrop reaches a maximum velocity as it falls towards the Earth.   [4]

> 1. Speed is constant[1], acceleration is zero[1].
> 2. The raindrop initially accelerates due to the force of gravity.[1] As the speed of the raindrop increases, the force upwards due to air resistance on the raindrop increases.[1] When the force due to gravity is balanced by the air resistance, the resultant force is zero[1] and terminal velocity is reached[1].

# VELOCITY-TIME GRAPHS

You can use a velocity-time graph to calculate acceleration and distance travelled.

## Calculating acceleration

The acceleration of an object can be determined from a velocity–time graph.
- A velocity-time graph shows how the velocity of an object changes over time
- The **gradient** of the graph equals the **acceleration** of the object

**Using the gradient**
- Gradient = $\Delta v$ / $\Delta t$
- The gradient equals acceleration because $a = \Delta v / t$

Deceleration is negative acceleration.

- The steeper the line, the greater the acceleration

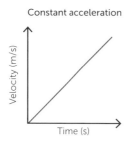
Constant acceleration

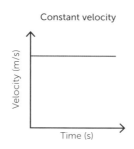
Constant velocity

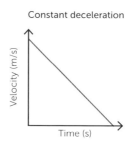
Constant deceleration

A straight line upwards represents **constant acceleration** as it has a positive gradient.
A horizontal line shows **constant velocity** and zero acceleration as the gradient is zero.
A straight line downwards represents **constant deceleration** as it has a negative gradient.

1.  A train starts from stationary and accelerates at a constant rate for 40 s. It then travels at a constant velocity of 30 m/s for 80s. It slows down to a stop with a constant deceleration for 20 s.
    Draw a velocity-time graph for this train journey. [3]

    1. *Line from (0,0) to (40,30).*[1] *Line from (40,30) to (120, 30).*[1] *Line from (120,30) to (140, 0).*[1]

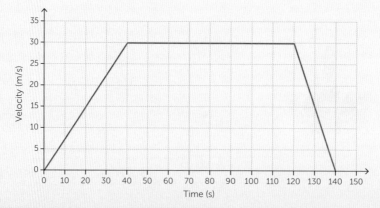

## Determining distance    Higher Tier

Velocity-time graphs can also be used to determine the **distance travelled** by (or displacement of) an object on a journey. The distance travelled over a period of time equals the **area under the velocity-time graph** for that period of time.

The area, and hence the distance travelled, can be calculated by using the values of time and velocity from the graph. Because the graph lines will be horizontal, or sloping up and down, the area under the graph will be a **triangle** or **rectangle** shape, or a combination.

**Area of a rectangle = width × height**

**Area of a triangle = ½ × base × height**

## Area under a distance-time graph

Calculate the distance travelled shown in the velocity-time graph by calculating the area under the graph.

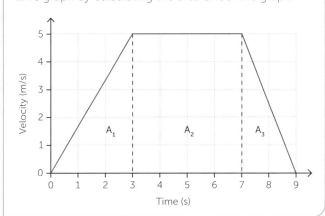

Check that the units match for both axes.

In this case area = s × m/s so the distance travelled is in m.

You could also calculate the distance by finding the area of a trapezium.

## Counting squares

You can also find the distance, or the area under the graph by **counting squares** under the graph line. This is straightforward when they are all whole squares, and more likely to be an estimate when there are several part squares under the line.

2. This is a velocity-time graph for a train journey. Calculate the total distance travelled by the train.    [1]

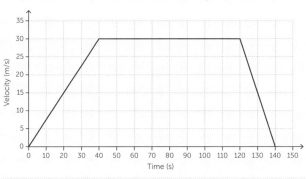

2.  distance = $(\frac{1}{2} \times 40 \times 30) + (80 \times 30) + (\frac{1}{2} \times 20 \times 30)$
    = 600 + 2400 + 300[1] = 3300 m[1]

# NEWTON'S FIRST LAW

**Sir Isaac Newton** was a mathematician and scientist who studied many things, including forces and motion. He developed three Laws of motion which are still used today to understand and describe forces and motion.

## The first law of motion

Newton's first Law states that:

- if there is **no resultant force** acting on a **stationary object**, then it will remain stationary.

- if there is **no resultant force** acting on a **moving object**, then the object continues to move at the same velocity (same speed, same direction).

A resultant force must act on an object for its velocity to change. When a resultant force acts on a stationary or moving object, the object will:

- stay at the same speed and change direction

- or change speed in the same direction

- or change both speed and direction.

This means when a vehicle travels at a constant speed, the driving force forward is balanced by the **resistive** forces of friction and air resistance acting against the vehicle. For the vehicle to speed up, slow down or change direction, the driving force, or the resistive forces, must change to produce a resultant force.

## Higher Tier only

The tendency of an object to stay at rest or continue moving with uniform motion is called **inertia**.

(a) This car is moving at a steady speed. Explain what will happen to the motion of the car. [2]

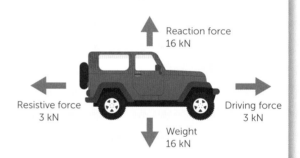

Reaction force
16 kN

Resistive force
3 kN

Driving force
3 kN

Weight
16 kN

(b) The driving force increases to 4 kN. Explain what happens to the car.  [2]

(a) *There is no resultant force on the car[1] so the car will continue moving at a steady speed[1].*

(b) *There is a resultant force on the car of 1 kN to the right[1] so the car accelerates to the right[1].*

# NEWTON'S SECOND LAW

An **estimate** is not an exact value. It is an approximate value.

## The second law of motion

Newton's second Law states that the acceleration of an object is:
- proportional to (∝) the resultant force acting on the object
- inversely proportional to the mass of the object.

You need to be able to recall and apply this equation.

The equation that links acceleration, force and mass is:

$$\text{resultant force} = \text{mass} \times \text{acceleration}$$

$F = m \times a$

$F$ = force in newtons, N

$m$ = mass in kilograms, kg

$a$ = acceleration in metres per second squared, m/s$^2$

## Estimating speed, acceleration and forces for road vehicles

Road vehicles undergo significant acceleration when overtaking or joining a motorway, for example. Typical masses are 30 000 kg for a lorry and 1500 kg for a car. If both vehicles accelerate from 15 m/s to 20 m/s then a greater force will be required to accelerate the lorry because it has a greater mass. Typical accelerations are ~0.4 m/s$^2$ for a lorry and ~2 m/s$^2$ for a car. The force required can then be estimated using $F = ma$.

So the typical forces are:

$F_{\text{lorry}}$ = 30 000 × 0.4 = 12 000 N          $F_{\text{car}}$ = 1500 × 2 = 3000 N

1. A car with a mass of 1200 kg accelerates at 3 m/s$^2$. Calculate the force needed to accelerate the car. [2]

2. A ball has a mass of 420 g and a resultant force of 5.1 N acting on it. Calculate the acceleration of the ball in m/s$^2$. [4]

3. **Higher only:** What is the inertial mass of the ball? [1]

## Inertial mass    Higher Tier

**Inertial mass** is:
- a measure of how difficult it is to change the velocity of an object
- defined as the **ratio** of force over acceleration $\dfrac{F}{a}$.

An object with a greater inertial mass needs a larger force to produce the same acceleration as an object with a smaller inertial mass.

1. $F = m \times a = 1200 \times 3^{[1]} = 3600 \ N^{[1]}$

2. $420 \ g = 0.42 \ kg^{[1]}$
   $F = m \times a$
   $5.1 = 0.42 \times a^{[1]}$
   $a = 5.1/0.42^{[1]}$
   $= 12.1 \ m/s^{2[1]}$

3. $0.42 \ kg.^{[1]}$

# REQUIRED PRACTICAL 7 (19)

## Investigating $F = ma$

This practical activity helps you to measure the effect on acceleration when force or mass are changed.

### Typical equipment

The weight of the hanging masses which is attached to the trolley by string going over the pulley provides the **force**.

The **mass** is changed by adding different masses to the trolley.

Each light gate records the speed of the trolley as the card passes through it. The two light gates record the time taken to travel between the light gates. The **acceleration** is then calculated from the change in speed and the time taken to travel between the light gates.

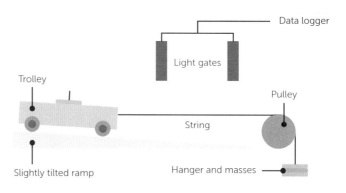

The string must be horizontal and in line with the trolley.

The hanging masses should just accelerate the car along the ramp – the number should be adjusted so that this happens.

### Alternative methods

- Instead of light gates, use a stopwatch to record the time taken for the trolley to travel equally measured distances on the ramp. You can also use ticker-tape and a ticker timer.

- Video the moving trolley so that the film can be slowed down and the distances and times measured more reliably.

- Replace the ramp and trolley with a linear air track and glider. This reduces the opposing friction force on the accelerating object.

## Varying the force

You can investigate the effect of varying the force on the acceleration of the trolley with a constant mass.

- The overall mass of the moving items (trolley, masses, string and hanger) must be kept constant.
- Masses are moved from the trolley and placed on the hanger.
- This increases the pulling force on the trolley but keeps the accelerating mass the same.

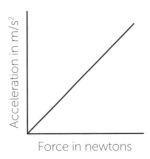

**a ∝ F** so a graph of acceleration against force should be a straight line through the origin.

## Varying the mass of an object

You can investigate the effect of varying the mass of an object on the acceleration of the trolley with a constant force.

- The same mass is kept on the hanger on the end of the string.
- The mass and the hangar on the string must give enough force to move the trolley at its greatest mass.
- Add different masses to the trolley.

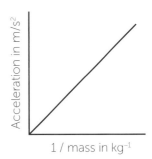

**a ∝ 1/m** so a graph of acceleration against 1/mass should be a straight line through the origin.

1. Explain why two light gates are used in the investigation. [2]
2. How do you calculate the force applied to the trolley? [1]
3. Explain why the ramp needs to be slightly tilted. [2]
4. Give **two** reasons why the length of the string used is important in this investigation. [2]
5. Describe the safety precautions you would need to take when carrying out this investigation. [2]

1. *The velocity of the trolley must be measured at two different points[1] so that acceleration can be calculated using $a = \Delta v/t$.[1]*
2. *Use $F = mg$[1]*
3. *The trolley will experience a friction force as it travels down the ramp[1] so tilting the ramp compensates for this slowing down of the trolley.[1]*
4. *The string needs to be long enough to allow the trolley to stay on the ramp as the hanging masses hit the floor[1] and short enough for the trolley to pass through both light gates before the masses hit the floor.[1]*
5. *Something soft, such as foam, is needed to stop the hanger and mass hitting the floor.[1] This is placed under the hanger.[1]*

# NEWTON'S THIRD LAW

## The third law of motion

Newton's third Law states that two objects exert **equal** and **opposite** forces on each other when they interact.

- The forces are sometimes called action-reaction forces.
- The forces act on different objects at the same time.
- Both forces are the same type of force e.g. push, pull.

> Be careful! Pairs of equal and opposite forces are not always examples of Newton's third law. The force of gravity acting downwards on a book and the reaction force of the table pushing up on the book, are both acting on the same object – the book. This is not an example of Newton's third law.

## Examples of Newton's third law

Newton's third law applies to both contact and non-contact forces. Here are some examples of Newton's third law in equilibrium situations. An equilibrium situation is when the forces are balanced. There is no resultant force in any direction.

**Non-contact forces:**

- The Earth exerts a gravitational force on the Moon, and the Moon exerts a gravitational force on the Earth.

**Contact forces:**

- A book exerts a pushing force on a table because of its weight, and the table exerts a pushing force back on the book which has equal value in the opposite direction.
- A person pushes against a wall with a force of 100 N. The wall pushed back against the person with a force of 100 N.

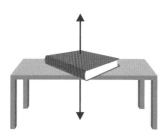

Explain why action-reaction forces are not the same as balanced forces.    [1]

*Balanced forces act on the same object.* [1] *Action-reaction forces act on different objects.* [1]

## Summary of Newton's laws

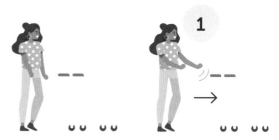

**1**

**Law of inertia:** A body will remain at rest of constant velocity unless acted on by an unbalanced force

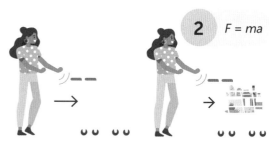

**2**    $F = ma$

**Law of force and acceleration:** The forces experienced by an object are proportional to its mass × the acceleration applied to it.

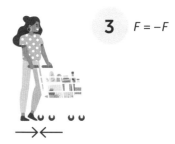

**3**    $F = -F$

**Law of action and reaction:** If two bodies exert a force on each other, the forces are equal in magnitude but opposite in direction.

# STOPPING DISTANCE

The stopping distance of a driven vehicle is the total distance that the vehicle travels between the point the driver decides to stop, and the vehicle actually stopping.

## Thinking distance and braking distance

The stopping distance of a vehicle is made up of two parts:

stopping distance = thinking distance + braking distance

The **thinking distance** is the distance the vehicle travels during the time the driver is reacting. It is linked to the driver's reaction time. The **braking distance** is the distance the vehicle travels after a force is applied on the brakes.

Given the same braking force, the stopping distance will increase as the speed increases. This is because the vehicle will travel further in the same amount of time.

## Stopping distance    Physics only

An emergency stop is when a driver brakes suddenly to try and stop a vehicle. This graph shows how stopping distance varies with speed for a typical car.

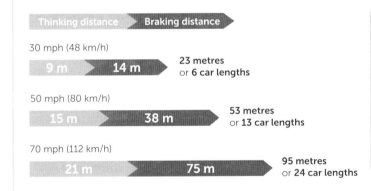

During braking, energy is transferred from the kinetic store of the car, which reduces to 0.

$E_k = mv^2$, so when $v$ doubles

$E_k = m \times (2v)^2 = m \times 4v^2 = 2v^2$

So, when the speed doubles, $E_k$ quadruples, and hence the braking distance is also 4 times greater.

The graphics above can be used to estimate how the distance travelled by a vehicle varies with speed when the driver makes an emergency stop.

---

1. Describe the relationship between the stopping distance and the speed of a vehicle. [1]
2. **Physics only:** Describe the relationship between thinking distance and speed. [1]
3. **Physics only:** Estimate the stopping distance for a typical car travelling at 18 m/s (40 mph). Use the graphic above to support your answer. [3]

1. *The greater the speed, the greater the stopping distance.[1]*
2. *Thinking distance is proportional to speed.[1]*
3. *38 m is halfway between 23m and 53m[1] but the distance will be less than that as the rate of increase of distance increases with speed[1] so an estimate is about 36 m[1].*

# REACTION TIME

**Reaction time** is the time a person takes to react to a situation.

## Factors affecting reaction time

Reaction times are typically in the range **0.2 s** to **0.9 s** and vary between different people.

Factors that affect reaction time include:

- when the driver is tired
- when the driver has taken alcohol or drugs
- when the driver is distracted such as using a mobile phone.

'Longer' can refer to time or distance so use terms carefully. Using 'greater' can be clearer.

## Measuring reaction time

Reaction times can be measured in different ways. Because the times involved are small, they are difficult to measure accurately. It is important to record lots of measurements, repeat them and calculate mean values.

Reaction times can be found by measuring the time between a person **detecting a stimulus** and then **responding** to it.

The stimulus could be a light or a sound, and the response can be recorded by pushing a button on a computer.

A typical value of reaction time is 0.7 s. Some people, such as sprinters, can improve their reaction times through training.

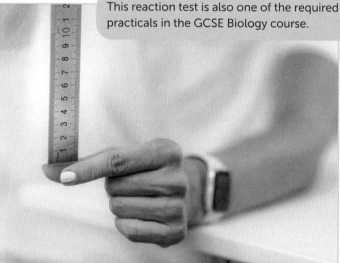

This reaction test is also one of the required practicals in the GCSE Biology course.

Reaction time can also be measured by catching a dropped ruler and comparing how far it falls for different people under different situations. The further the ruler falls, the greater the reaction time of the person.

---

1. Explain the relationship between reaction time and thinking distance. [2]
2. Give **two** disadvantages of using the dropped ruler method to measure reaction time compared with a computer. [1]
3. Suggest why there is a legal limit for alcohol levels when driving. [2]

   1. *The thinking distance is the distance the car travels while the driver reacts to a danger.[1]*
      *If the driver has a longer reaction time, the car will have travelled further – the thinking distance is greater.[1]*

   2. *The ruler may be dropped from a slightly different height each time due to human error.[1]*
      *The time may also be affected by the reaction time of the person using the stopwatch.*

   3. *Alcohol can lengthen a person's reaction time[1] which increases the overall stopping distance of a vehicle[1].*

# FACTORS AFFECTING BRAKING DISTANCE 1

The **braking distance** of a vehicle is related to the **condition of the vehicle**, the **surface** being driven on and the **weather**.

## Braking distance

Wet or icy conditions, and worn tyres reduce friction between the tyres and the road surface. Less friction between the tyres and the surface mean that the vehicle will travel further whilst braking.

Worn brakes require a larger force to be applied so will also increase the braking distance.

The graph below shows values for the stopping distance of a typical car in good condition on dry roads. It can be seen that as speed increases, braking distance has a greater impact than thinking distance on the overall stopping distance.

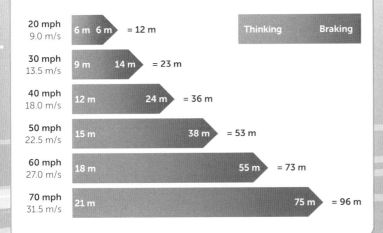

1. A car is travelling at 30 mph in icy conditions.

   (a) Use the graphic opposite to estimate the stopping distance. Tick **one** box.  [1]

   ☐ 14 m  ☐ 20 m
   ☐ 23 m  ☐ 28 m

   (b) Explain your answer to part (a).  [2]

2. Explain why worn tyres increase braking distance but not thinking distance.  [2]

*1. (a) 28m.[1]*

   *(b) Icy conditions increase the braking distance[1] so the overall stopping distance will be higher than 23 m[1].*

*2. The thinking distance is the distance travelled as the driver is responding to a stimulus to stop the vehicle.[1] The condition of the tyres has no impact on the driver's reaction as they are not braking at this time.[1] Less grip on the tyres results in less efficient braking, increasing braking distance.[1]*

# FACTORS AFFECTING BRAKING DISTANCE 2

A moving vehicle has **kinetic energy** which is dependent on velocity ($E_k = \frac{1}{2}m v^2$). See also **page 71**.

## Braking

When a vehicle stops, the energy from the kinetic energy store of the vehicle is transferred to other energy stores. When a force is applied to the brakes of a vehicle, the brakes press against the wheel.

Work is done by the friction force between the brakes and the wheel. Energy is transferred mechanically from the kinetic energy store of the vehicle to the **thermal energy** store of the brakes. This causes the vehicle to slow down and the temperature of the brakes to increase.

The greater the speed of a vehicle, the greater the **braking force** needed to stop the vehicle in a given distance. Greater braking forces cause greater **decelerations**.

A higher speed means that there is more energy in the kinetic energy store. As this energy is transferred to the thermal energy store of the brakes, the brakes can overheat and stop working properly. This can cause the driver to lose control of the vehicle.

Large decelerations can also cause injuries to the occupants of the vehicle.

1. Describe the braking force needed to stop a vehicle in the same distance when the speed doubles. [2]

2. **Higher only:** A driver of a lorry with a mass of 36 000 kg brakes suddenly. The lorry decelerates from 15 m/s to 0 m/s in 20 s. Estimate the braking force on the lorry. [2]

3. **Higher only:** Estimate the force needed to stop a car with a mass of 1500 kg travelling at 60 mph. Use the graphic on **page 97**. [3]

   1. *If the speed of the vehicle doubles, the kinetic energy quadruples,[1] so the force needed to stop the vehicle in the same distance quadruples[1].*

   2. *a = Δv/t = (15 - 0) / 20 = 0.75 m/s²[1]*

      *F = m × a = 36 000 × 0.75 = 27 000 N.[1]*

   3. *Stopping distance = 55 m; speed = 27 m/s*
      *Fd = $\frac{1}{2}mv^2$*
      *F × 55 = 0.5 × 1500 × 27²*
      *F = (0.5 × 1500 × 27²) / 55*
      *= 9900 N[1]*

## Estimating braking forces    Higher only

Braking forces for vehicles under typical conditions can be estimated when the mass and deceleration of the vehicle are known.

Deceleration can be calculated using $a = \dfrac{\Delta v}{t}$ and the force can be calculated using $F = ma$.

When you know the stopping distance, you can also calculate the force using:

*work done (F d) = kinetic energy of car ($\frac{1}{2}mv^2$)*

# CONSERVATION OF MOMENTUM

A moving mass with a velocity also has **momentum**.

## Calculating momentum

Note that *p* is also the symbol used for pressure.

Momentum is a **vector** quantity. A stationary object has zero momentum because v = 0. Momentum is defined by the equation:

momentum = mass × velocity

$p = m\,v$

$p$ = momentum in kilograms metre per second, kg m/s

$m$ = mass in kilograms, kg. $v$ = velocity in metres per second, m/s

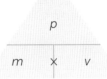

A bird with a mass of 250 g flies at a velocity of 18 m/s. Calculate the momentum of the bird. [3]

$250\ g = 0.25\ kg^{[1]}$     $p = m \times v = 0.25 \times 18^{[1]} = 4.5\ kg\ m/s.^{[1]}$

## Conservation of momentum

The law of conservation of momentum states that in an event in a **closed system** such as a collision, the total momentum before the event equals the total momentum after the event. A closed system is where no energy escapes or enters from outside the system, and no external forces are acting on the system.

Momentum can have positive and negative values depending on the direction the object moves.

| Before collision | Collision | After collision |
|---|---|---|
| a) Red ball has momentum *p*. Blue ball is stationary. | | Red ball stops. Blue ball now moves in the same direction with the same momentum *p* that the red ball had. |
| b) Red ball has velocity *v*. Blue ball is moving in the opposite direction with velocity −*v* | | Both balls continue to move but in the opposite direction, so red ball now has velocity −*v*, blue ball velocity *v*. |
| c) Red ball has velocity *v*. Blue ball moving in the same direction but with a lower velocity. | | Red ball velocity decreases. Blue ball velocity increases. |
| d) Red ball has momentum *p*. Blue ball is stationary. | | Both balls stick together and move with the same velocity. |

# CONSERVATION OF MOMENTUM CALCULATIONS

When the momentum before or after an event is known, along with some of the masses and velocities, an unknown value for mass or velocity can be calculated using the conservation of momentum.

## Collision between two objects

Consider a collision between two objects. When the two objects remain separate, then their individual momentums before and after the collision can be calculated when you know the velocity of one of the objects after the collision.

> Check the values you calculate for unknown mass and velocity make sense in terms of the known values.

Before collision            After collision

$m_1v_1$     $m_2v_2$     $m_1v_3$     $m_2v_4$

$$m_1v_1 + m_2v_2 = m_1v_3 + m_2v_4$$

If two objects stick together after a collision, then the new mass is the total of the two colliding masses.

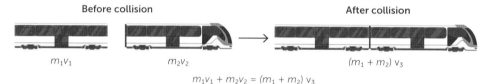

Before collision            After collision

$m_1v_1$     $m_2v_2$     $(m_1 + m_2)\,v_3$

$$m_1v_1 + m_2v_2 = (m_1 + m_2)\,v_3$$

Write out the equation in full for the momentum before and after to isolate and calculate the unknown value.

1. A supermarket trolley with a mass of 14.0 kg moves at a velocity of 2.1 m/s. It hits an identical trolley moving at a velocity of 1.5 m/s in the same direction. The two trolleys move off together. Calculate the velocity of the two trolleys as they move together. [3]

2. Ice skater 1 has a mass of 60 kg and skates in a straight line with a velocity of 3 m/s. Ice skater 1 bumps into stationary ice skater 2. Ice skater 2 moves forward with a velocity of 2 m/s. Ice skater 1 continues to move forward with a velocity of 0.5 m/s. Calculate the mass of ice skater 2. [3]

1.   $m_1v_1 + m_2v_2 = (m_1 + m_2)v_3$    $(14 \times 2.1) + (14 \times 1.5)$
     $= (14 + 14)v_3$[1]    $v = (29.4 + 21) / 28$[1]
     $= 1.8$ m/s[1]

2.   $m_1v_1 + m_2v_2 = m_1v_3 + m_2v_4$    $(60 \times 3) + (75 \times 0)$
     $= (60 \times 0.5) + (m \times 2)$[1]    $v = (180 - 30) / 2$[1]
     $= 75$ kg[1]

# CHANGES IN MOMENTUM

A **change in momentum** occurs when a force acts on a moving object, or on a stationary object that is able to move.

## Force and momentum

By combining Newton's second Law with the equation used for calculating acceleration, it can be shown that force is the rate of change in momentum.

We can combine the equations $F = ma$ and $a = \dfrac{\Delta v}{t}$ to give the equation:

$$F = \dfrac{m\,\Delta v}{\Delta t}$$

$F$ = force in newtons, N

$m\Delta v$ = change in momentum in kilograms metre per second, kg m/s

$\Delta t$ = change in time in seconds, s

> You need to be able to recall and apply this equation.

Rapid changes in momentum can be very dangerous. If a person decelerates over a very short space of time (e.g. travelling fast and stopping suddenly) a large force is exerted.

If the change of momentum can be made to occur over a longer period of time, then this slows down the change in momentum and the force is reduced.

**Safety features** such as airbags, crumple zones in cars, cycle helmets, cushioned surfaces for playgrounds, gymnasium crash mats and seat belts increase the time taken to bring the body to rest, so reducing the force needed to stop the body. They all reduce the rate of change in momentum hence reduce the force and the risk of injury.

> $m\Delta v = mv - mu$ = change in momentum during an event such as a collision.

1. A 65 kg cyclist travels at 6 m/s. The cyclist comes to a stop in 1.5 s. Calculate the force experienced by the cyclist.    [2]

2. Explain how a gymnasium crash mat reduces the risk of injury for the gymnast.    [4]

3. An object with a mass of 0.4 kg hits the ground with an acceleration of 12.5 m/s². The object hits the ground with velocity $v$ and comes to a stop in 0.1 s. Calculate $v$ in m/s.  [3]

1. $F = m\Delta v / \Delta t$
   $= (65 \times 6) / 1.5^{[1]} = 260\ N^{[1]}$

2. When a gymnast hits the floor they decelerate rapidly as their velocity goes from a high value to zero in a short time[1] so there is a high force on the gymnast[1]. The gym mat increases the time taken for the change in momentum to occur.[1] This reduces the force (and risk of injury) experienced by the gymnast.[1]

3. $F = m\Delta v / \Delta t = 0.4 \times 12.5$
   $= 0.4\ (v - 0)/0.1^{[1]}$
   $v = 12.5 \times 0.1^{[1]} = 1.25\ m/s^{[1]}$

# EXAMINATION PRACTICE

01 This question is about forces.

01.1 Which quantity is a vector quantity? Tick **one** box. [1]

☐ displacement ☐ distance ☐ mass ☐ time

The diagram shows the forces on a box, $F_x$ and $F_y$.

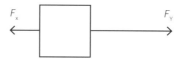

01.2 Compare the two forces. [2]

01.3 Explain what will happen to the box. [2]

01.4 **Higher Tier only:** Explain why momentum is a vector quantity. [2]

01.5 Give **two** examples of non-contact forces. [2]

01.6 Define weight. [1]

01.7 Calculate the weight of an object with a mass of 1.6 kg. Use $g$ = 9.8 N/kg [2]

01.8 Name an instrument used to measure weight. [1]

02 This question is about resultant forces.

The diagram shows the forces acting on a block.

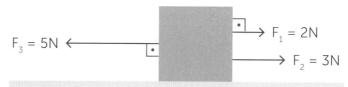

02.1 Calculate the resultant force on the block. [2]

02.2 **Higher Tier only:** Draw a free body diagram of a cyclist moving to the right along a flat road. Label all four forces acting on the cyclist. [3]

02.3 **Higher Tier only:** A single force of 3 N acts at 48° to the vertical. Draw a scale diagram to resolve the force into two forces. Use graph paper. [2]

02.4 **Higher Tier only:** The diagram represents two forces on a boat. The forces are acting at right angles to each other. Draw a scale diagram to determine the resultant force on the boat. [3]

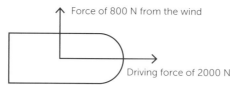

03 This question is about the energy transfer when a force does work.

03.1 Give the equation that links distance, force and work done. [1]

03.2 Show how the work done in raising a mass through a vertical distance is equal to the energy transferred to the gravitational potential energy store of the mass. [4]

03.3 The work done to stop a car on a flat surface is 6600 J. Calculate the distance travelled, in m, when the braking force is 1200 N. [3]

03.4 Convert 500 Nm into kilojoules (kJ). [2]

04 This question is about elasticity.

04.1 Write the equation that links force, extension and spring constant. [1]

04.2 A spring stretches 0.07 m when a force of 2.1 N is applied. The deformation is elastic. Calculate the spring constant of the spring. Give the unit. [4]

04.3 The sketch shows a small toy that works by compressing a spring.
The compression of the spring is 3.5 cm.
The spring constant is 500 N/m.

Calculate the energy stored in the spring. [3]

04.4 A student investigates the relationship between force and extension for a spring. The results are shown in the table.

| Mass in g | Force in newtons | Length of spring in cm | Extension in cm |
|-----------|------------------|------------------------|-----------------|
| 0 | 0 | X | 0 |
| 10 | 0.1 | 2.4 | 0.2 |
| 20 | 0.2 | 2.6 | 0.4 |
| 30 | 0.3 | 2.8 | 0.6 |
| 40 | 0.4 | 3.0 | 0.8 |
| 50 | 0.5 | 3.2 | 1.0 |

Calculate X, the original length of the spring. [1]

04.5 The diagram shows a graph of extension in cm against force in newtons for another spring, Spring A.
Plot a graph of extension against force on the axes above for the spring in the table.
Draw a line of best fit. [3]

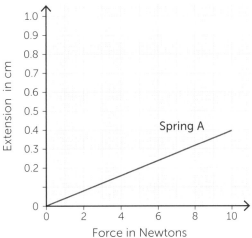

04.6 Determine the spring constant for Spring A. [2]

**Physics only**

05    This question is about rotational forces.

05.1  Calculate the moment on the nut.                                              [2]

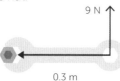

9 N

0.3 m

The figure below shows a hanging mobile. The mobile is balanced.

05.2  Describe what balanced means in terms of turning forces.                       [1]

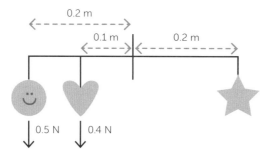

0.2 m

0.1 m        0.2 m

0.5 N   0.4 N

05.3  Calculate the weight of the star.                                              [4]

**Physics only**

06    This question is about pressure in fluids.

A fluid exerts a force of 25 N on a surface with an area of 0.2 m².

06.1  Calculate the pressure the fluid exerts on the surface. Include the unit.      [3]

06.2  Describe what causes atmospheric pressure.                                     [2]

**Higher Tier only:** A diver swims horizontally under water.
The diver experiences a pressure of 29 400 Pa.

06.3  Calculate the depth of the diver in the water.
       Density of water is 1000 kg/m³. Use $g$ = 9.8 N/kg.                           [3]

06.4  Explain why the diver experiences an upthrust.                                 [3]

07    This question is about motion.

07.1  The diagram shows the journey of a car from Town A to Town B
       Determine the displacement of the car.                                        [2]

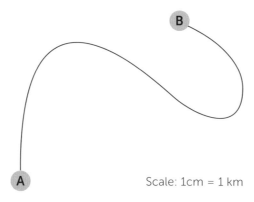

B

A          Scale: 1cm = 1 km

07.2 The car travels a further 3.6 km to get home. This journey takes 150 s. Calculate the mean speed of the car for this journey in m/s. [4]

The diagram shows a distance-time graph for the journey of an object.

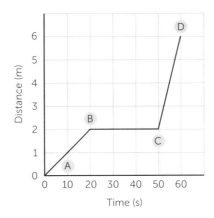

07.3 An object travels 2 m in 20 s.
The object is then stationary for 30 s
The object then travels 4 m in 10 s.
Draw a distance-time graph for the motion of the object. [3]

07.4 Calculate the mean speed of the object for the whole journey. [3]

07.5 **Higher Tier only:** Sketch a distance-time graph for a decelerating object. [1]

08   This question is about acceleration.

08.1 A car accelerates at 1.5 m/s² for 8s.
Calculate the change in velocity in m/s. [3]

08.2 A cyclist increases speed from 5.0 m/s to 9.0 m/s
with an acceleration of 0.5 m/s².
Calculate the distance travelled by the cyclist. [3]

08.3 The diagram shows a velocity-time graph for three
objects, A, B and C.
Describe the motion of the three objects. [3]

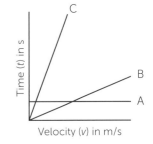

08.4 The diagram below shows a velocity-time graph for the journey of an object.
Calculate the acceleration of the object for the first 20s of the journey. [2]

08.5 **Higher Tier only:** Calculate the total distance travelled by the object during the journey. [2]

08.6 **Physics only:** Sketch a velocity-time graph for an object that reaches terminal velocity. Label the terminal velocity on your graph. [3]

09 This question is about Newton's Laws of motion

09.1 **Higher Tier only:** What is the tendency of an object to stay at rest or continue moving with uniform motion called? Tick **one** box. [1]

☐ Acceleration          ☐ Deceleration          ☐ Inertia          ☐ Momentum

09.2 What must happen to make a stationary object move? [1]

09.3 The diagram shows an object moving to the right at a constant speed.

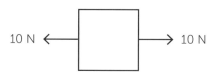

10 N ←          → 10 N

Explain what happens to the movement of the object when an upward force of 10 N also acts on the object. [2]

09.4 A resultant force of 250 N acts on an object. The object accelerates at 12.5 m/s$^2$. Calculate the mass of the object in kg. [3]

09.5 Explain the motion of this box. [2]

4 kg    → F = 20N

A student investigates the effect of changing mass on the acceleration of a trolley using this apparatus.

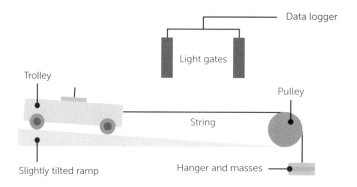

09.6 Explain why the number of masses on the hanger is kept constant. [2]

09.7 Give **one** safety factor to consider in this investigation. [1]

09.8 Define Newton's third law of motion. [1]

10    This question is about forces and braking vehicles.

    10.1    Define the stopping distance of a vehicle.                                                          [1]

The graphic below shows some typical stopping distances for a car driving at different speeds.

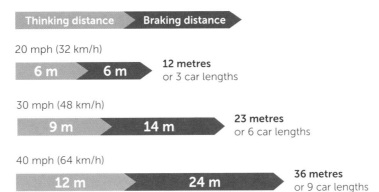

    10.2    **Physics only:** Compare the stopping distances for the car at 20 mph and 40 mph.              [2]

    10.3    Which of these is correct when a driver is tired? Tick **one** box.                          [1]
        ☐ Braking distance increases
        ☐ Reaction time increases
        ☐ Stopping distance decreases
        ☐ Thinking distance decreases

    10.4    Give **two** factors that increase the braking distance of a vehicle.                       [2]

    10.5    Explain how a large deceleration can cause brakes to overheat in a vehicle.               [3]

11    **Higher Tier only:** This question is about momentum.

    11.1    Calculate the momentum of a 5.2 kg ball rolling with a velocity of 0.3 m/s.                 [2]

    11.2    Give the law of conservation of momentum.                                                    [1]

    11.3    **Physics only:** Train carriage 1 has a mass of 15 000 kg. Train carriage 1 moves at a velocity of 8.0 m/s and hits train carriage 2 which is moving at a velocity of 3.0 m/s in the same direction. The two carriages move off together at a velocity of 6.0 m/s.
        Calculate the mass of carriage 2.                                                           [3]

    11.4    **Physics only:** A 2.5 kg stone hits a wall at 15 m/s. The stone comes to a stop in 0.2 s. Calculate the force experienced by the stone.                                                   [2]

# TRANSVERSE AND LONGITUDINAL WAVES

A **wave** is a **vibration (oscillation)** about an undisturbed position. Waves transfer **energy** and information from one place to another. They do not transfer **matter**.

**Mechanical** waves are oscillations (vibrations) of particles and transfer energy through a medium. **Electromagnetic** waves are oscillations of electrical and magnetic fields and also transfer energy. Waves can be **transverse** or **longitudinal**.

## Transverse waves

Perpendicular means at right angles.

**Water waves**, or ripples, are an example of a transverse wave. The water particles vibrate up and down about a fixed point. The vibrations are **perpendicular** to the direction of energy transfer. It is the wave and not the water particles that travel.

Parallel means in the same direction.

**Sound waves** travelling through air are an example of a longitudinal wave. Air particles vibrate backwards and forwards about a fixed point. The vibration is **parallel** to the direction of energy transfer. Again, it is the wave not the air particles that travel.

Longitudinal waves show areas of **compression** and **rarefaction**. A compression is where the particles are closer together. A rarefaction is where the particles are further apart.

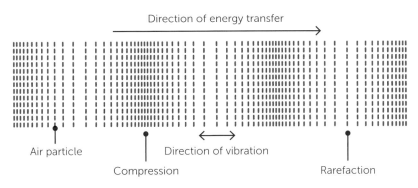

Direction of energy transfer

Air particle · Direction of vibration · Compression · Rarefaction

1. Give **one** similarity and one difference between a longitudinal and a transverse wave. [2]
2. Describe how a slinky spring can be used to show that for both types of wave the particles vibrate about a fixed position. [3]

1. *Similarity: both transfer energy / both are oscillations; about an undisturbed position.[1] Difference, any one from: Vibration is perpendicular to the direction of travel for transverse waves and parallel to the direction of travel for longitudinal waves / all longitudinal waves need a medium to travel through, some transverse waves do not (EM waves).[1]*

2. *Mark a point on the slinky and fix it at one end.[1] For longitudinal waves, move the other end back and forth to show that the point also moves back and forth but does not move forward.[1] For the transverse wave, move the end of the slinky up and down to show that the point also moves up and down but does not move forward.[1]*

# PROPERTIES OF WAVES

You can describe wave motion in terms of the wavelength, amplitude, frequency and period of the wave.

### Amplitude and wavelength

The **amplitude** of a wave is the maximum displacement of a point on a wave away from its **undisturbed position**. The undisturbed position is where there is no vibration. It is also known as the rest position, or the equilibrium position.

> Use a ruler when drawing lines to show the amplitude and wavelength on a wave.

The **wavelength (λ)** of a wave is the distance from a point on one wave to the equivalent point on the adjacent wave.

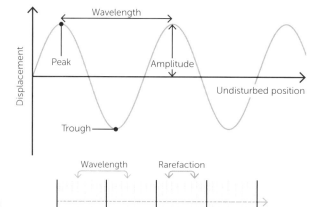

Transverse wave:

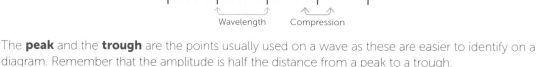

Longitudinal wave:

The **peak** and the **trough** are the points usually used on a wave as these are easier to identify on a diagram. Remember that the amplitude is half the distance from a peak to a trough.

In a longitudinal wave, λ is the distance between a compression and a compression, or a rarefaction and a rarefaction.

It is difficult to show amplitude on a longitudinal wave so amplitude is usually shown on a transverse wave.

### Frequency

The **frequency (f)** of a wave is the number of waves passing a point each second.

The **period (T)** of a wave is the time taken for one full cycle of the wave.

$$\text{period} = \frac{1}{\text{frequency}}$$

$$T = \frac{1}{f}$$

$T$ = period in seconds, s

$f$ = frequency in hertz, Hz. **1 Hz = 1 wave per second.**

> You need to be able to select this equation from the equation sheet and apply it

## Wave speed

The **wave speed (v)** is the speed at which the energy is transferred, or the speed that the wave moves, through a medium.

All waves obey the **wave equation**:

$$\text{wave speed} = \text{frequency} \times \text{wavelength}$$

$v = f\lambda$

$v$ = wave speed in metres per second, m/s

$f$ = frequency in hertz, Hz

$\lambda$ = wavelength in metres, m

You need to be able to recall and apply this equation.

A **medium** is a something that a wave can be transmitted through.

## Transmission of sound waves     Physics only

Sound waves travel at different speeds in different materials. When a sound wave travels from one medium, such as air, to another medium, such as water, the velocity changes. The frequency of the wave does not change as it depends on the source of the vibrations, not the medium it travels through.

The wave equation tells us that velocity is proportional to wavelength so if the velocity changes, so too will the wavelength. A change in velocity can also mean a change in direction.

Temperature and density can affect the speed of sound.

1. Label the amplitude and the wavelength on this diagram of a wave. [2]
2. Describe the relationship between wavelength and frequency.     [1]
3. Calculate the period of a wave with a frequency of 0.5 Hz.     [2]
4. A wave travels with a frequency of 7.0 Hz and a wavelength of 0.4 m. Calculate the speed of the wave.     [2]

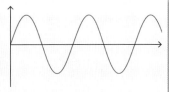

1.  *Labelled: Amplitude as the vertical height from horizontal line to peak.[1] Wavelength as the distance between any point and the same point in the next wave.[1]*
2.  *The longer the wavelength, the lower the frequency.[1]*
3.  *T = 1/f = 1 / 0.5[1] = 2s[1]*
4.  *v = f$\lambda$ = 7.0 × 0.4[1] = 2.8 m/s[1]*

# MEASURING WAVE SPEED

There are various methods for measuring the speed of sound in air and the speed of ripples on the surface of water.

## Measuring the speed of sound in air

One possible method is:

- A person makes a loud noise (e.g. banging together blocks of wood) and indicates as they make the sound.

- A second person stands 100–200 m away and starts a stopwatch when they see the visual sound indication.

- They stop the watch when they hear the sound and record the time taken.

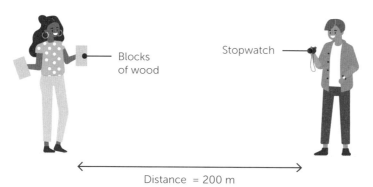

Blocks of wood

Stopwatch

Distance = 200 m

> The sound is seen being made before it is heard because light travels much faster than sound in air (3 × 10⁸ m/s, and 330 m/s)

An alternative method is to record the time taken to hear an echo from a wall (the sound travels to the wall and back). This has the advantage of a greater distance, so the time taken for the sound wave to travel will be longer and easier to measure. Also, the person with the stopwatch stands next to the person making the sound which may mean the stopwatch is started more accurately.

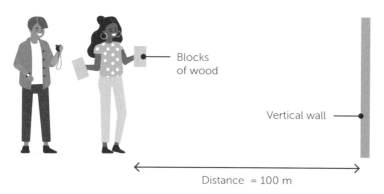

Blocks of wood

Vertical wall

Distance = 100 m

## Measuring the speed of waves and ripples on a water surface

One possible method is to time how long it takes a water wave to travel a known distance such as along a sea wall. A ripple tank can also be used for this.
See the next page for more details. **RPA 8 (20)**.

1. Explain how the reaction time of the person timing affects the measurement of the speed of sound using a wall. [2]

2. You are going to measure the speed of a water wave by timing how long it takes to travel a known distance along a sea wall.
   Describe how you would take your measurements. [3]

   1. *The reaction time of the person timing introduces an error which makes the time longer than the true value.[1] This has the effect of calculating the speed to be lower than its true value.[1]*

   2. *Identify two obvious features or points on the sea wall that you can use for timing the waves.[1] Measure the distance between these two points.[1] Time how long it takes several waves to travel between these two points.[1]*

You can see a lightning flash before you hear the clap of thunder. The light from the flash travels at the speed of light, whereas the sound from the clap of thunder travels at the speed of sound. You can use this information to calculate how far away the lightning flash was.

## Investigating waves

This practical activity helps you make observations of waves in fluids and solids to identify the suitability of apparatus to measure speed, frequency and wavelength.

Both investigations involve measurements of wavelength, and measurements or recording of frequency. You can calculate speed using: **wave speed = frequency × wavelength**.

### Water waves in a ripple tank

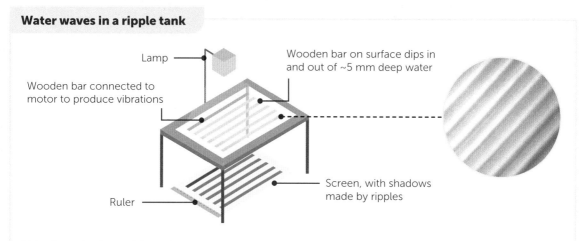

Lamp

Wooden bar on surface dips in and out of ~5 mm deep water

Wooden bar connected to motor to produce vibrations

Screen, with shadows made by ripples

Ruler

Light from the lamp shines through the water and an image of the waves can be observed on white card below the tank. The image will look clearer in a darker room.

**To measure wavelength:** lay a metre ruler at right angles to the waves. Measure the total length the waves travel. Divide the total length by the number of waves to find the wavelength.

**To measure frequency:** set the vibration generator to a low frequency so it is possible to count the number of waves passing a point in 10 seconds. Record observations from as many people as possible and calculate a mean.

The **number of waves** is needed for both values and can be difficult to measure. The waves can be video recorded and slowed down, or paused, so that the number of waves can be measured more accurately. The timer must also be recorded so that the real time taken is known to determine the frequency.

1. Explain how the frequency is calculated from the number of waves passing in 10 seconds. [1]
2. 18 waves pass a point in 10 seconds. Calculate the frequency of the waves. [1]
3. The length of the tank is 27 cm. 9 waves are seen on the tank.
   Calculate the wavelength of the waves [2]
4. Calculate the speed of the waves in the ripple tank. [2]

   1. *Frequency = number of waves / 10.[1]*
   2. *Frequency = 18/10 = 1.8 Hz[1]*
   3. *2.7 cm = 0.27 m[1] wavelength = 0.27 / 9 = 0.03 m[2]*
   4. *v = 1.8 × 0.03[1] = 0.054 m/s[1]*

## Waves on a stretched spring

When a string is vibrated at certain frequencies, **standing waves** are generated on the string. This is a wave that appears to be stationary so it is easier to measure the wavelength than from a travelling wave.

A standing wave is generated on a string by vibrating the string at one end. The tension in the string is adjusted to get a visible standing wave.

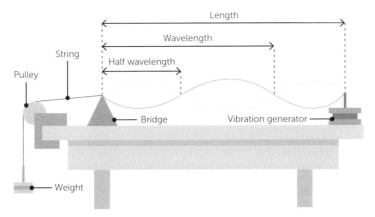

**Wavelength:** Use a metre ruler to measure the total length of the standing wave produced. Count the number of **half wavelengths** and determine the wavelength. Vary the power supply to the vibration generator to produce different wave frequencies. Measure the wavelengths for standing waves with different numbers of half wavelengths.

$$wavelength = \frac{total\ length}{number\ of\ half\ wavelengths} \times 2$$

**Frequency:** This is the frequency supplied to the vibration generator so it is recorded rather than measured.

5. What is the purpose of the hanging mass? [1]
6. Calculate the wavelength of this standing wave in m. [3]

   5. *To keep the string tight.[1]*
   6. *21 cm = 0.21 m[1] 0.21 / 3 × 2[1] = 0.14 m[1]*

21.0 cm

# REFLECTION OF WAVES

When waves move between two different materials, different phenomenon occur at the boundary between the two materials.

## Boundary between two different materials

Waves can be **reflected**, **absorbed** or **transmitted** at the boundary between two different materials, such as air and water. This happens to both transverse and longitudinal waves.

The materials involved, the wavelength and the angle of the incident wave determine what happens at the boundary.

When waves are absorbed, the energy of the waves is transferred to the material and the wave stops.

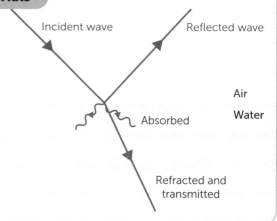

## Reflection

The reflection of a sound wave is called an echo.

The incident wave meets the surface at the angle of incidence and a reflected wave is reflected off the surface from this point.

The **normal** is a line at 90° to the surface from where the angles are measured.

The **angle of incidence** (*i*) is the angle between the incident wave and the normal.

The **angle of reflection** (*r*) is the angle between the reflected wave and the normal.

For all waves, the **angle of incidence = the angle of reflection.**

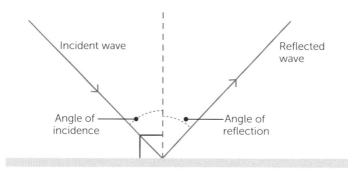

## Ray diagrams

You can draw a ray diagram to show the reflection of a wave such as light at a surface which is the boundary between two materials. When light waves, or **rays**, are reflected by a mirror, the **image** of the object appears to be behind the mirror.

Use a protractor, sharp pencil and ruler for drawing ray diagrams.

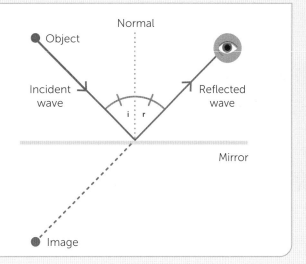

## Refraction

**Refraction** occurs when light is transmitted at a boundary between two transparent materials. The direction of the wave changes. When going from air to glass or water, the ray bends towards the normal.

Incident waves that are travelling along the normal and are transmitted at the surface will pass straight through the material without changing direction. The waves are at 90° to the surface

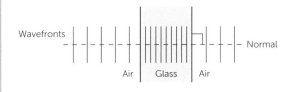

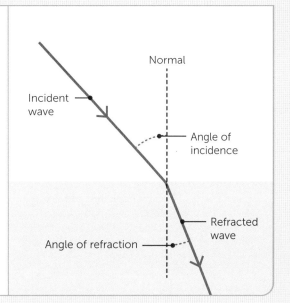

1. What happens to an incident wave that is at 90° to a surface and is transmitted by it?
   A. The wave cannot pass through the material.
   B. The wave continues through the material in a different direction.
   C. The wave continues through the material in the same direction.
   D. The wave is reflected back in the opposite direction. [1]

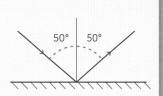

2. Draw a ray diagram to show the path of a reflected wave with an angle of incidence of 50°. [3]

   1. *C. The wave continues through the material in the same direction.[1]*
   2. *Normal line.[1] Angle of incidence = 50°.[1] Angle of reflection = 50°.[1]*

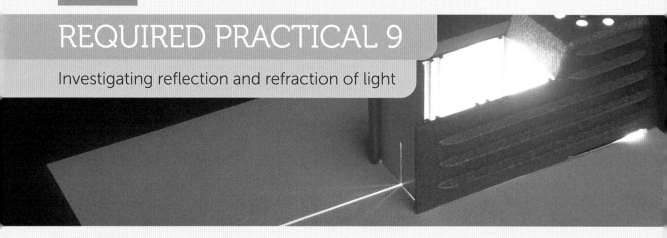

# REQUIRED PRACTICAL 9

## Investigating reflection and refraction of light

**This practical activity helps you to observe light interacting with matter.**

### Observing light behaviour

A narrow ray of light is used for this investigation. This ray can be produced by using a ray box with a lens and a single slit. A light ray box can get hot and should be switched off between readings to allow it to cool down.

Record the path of the ray of light and measure the angles of incidence, reflection and refraction.

There are two parts to this practical:
• Reflection of light by different types of surface
• Refraction of light by different substances.

Reflection and refraction can be observed at the same time when **transparent blocks** of different materials are used as both the surface and the substance. Glass and Perspex are common materials to use. Water in a straight sided tank is another option.

Use a protractor to measure the angle between the ray of light and the normal line. Light travels in straight lines so use a ruler for drawing light rays.

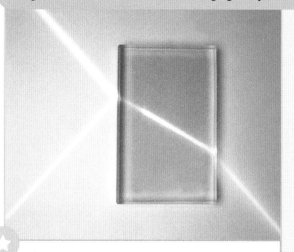

Remember to add an arrowhead to each ray to show the direction the light is travelling.

### Basic method

The basic method is as follows:
• Draw around a transparent block on a large sheet of white paper.
• Add a normal line in the centre of one side of the block.
• Shine a light ray at the point the normal meets the block.
• Adjust the angle until a reflected ray and a refracted ray are seen.
• Draw at least two small crosses along each ray so they can be joined up.
• Remove the ray box and block.
• Join the crosses to show the paths of the incident, reflected and refracted rays.
• Measure the angles of incidence, reflection and refraction.
• Repeat for different transparent blocks.

## Observations

Diagrams for each substance should look similar to this.

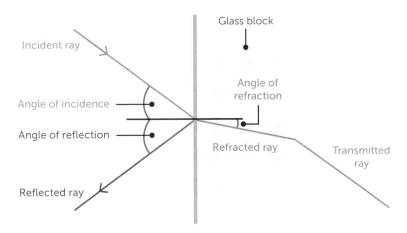

The results of the angles for the different materials allows the following conclusions to be made.
- For all reflections at different surfaces, angle of incidence = the angle of reflection.
- For refraction, the angle of refraction depends on the substance.

1. Explain the advantage of using a narrow ray of light. [2]
2. Describe how the rays of light can be seen more clearly. [1]
3. What happens to the angle of incidence and the angle of reflection when the direction of the incident ray is reversed? [1]
4. Explain why a mirrored surface cannot be used for observing the refraction of light. [2]

   1. *A narrow ray reduces errors[1] as the path of light can be traced/angles can be measured more accurately.[1]*
   2. *Carry out the investigation in a darkened room.[1]*
   3. *Nothing, because the angle of incidence still equals the angle of reflection.[1]*
   4. *Light is not transmitted by a mirrored surface[1] it is all reflected[1].*

# SOUND WAVES

Sound waves are longitudinal waves that transfer energy by causing vibrations. They are produced when objects vibrate.

## Travelling waves

Sound waves need a medium, such as a solid, liquid or gas, to travel through. They cannot travel in a vacuum.

Wave speed changes when waves move from one medium to another. Sound travels faster in a solid than in a gas because the particles are much closer together so they can pass on the vibrations more easily.

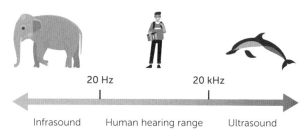

Sound waves enter the **ear** and cause a thin membrane called the **ear drum** to vibrate. This then passes on the vibration mechanically to small bones in the inner ear mechanically. It is then passed to the brain as an electrical signal.

## Human hearing range

**Human hearing** is limited by the frequencies that will cause the ear drum to vibrate. The range of normal human hearing is **20 Hz** (low pitch) to **20 000 Hz** (high pitch). The range decreases as people get older, and regular exposure to loud noises can damage the eardrum so it vibrates less effectively.

## Microphones

A **microphone** works in a similar way to the ear. The vibrations of the sound waves make a thin paper sheet called a diaphragm vibrate. The diaphragm is attached to a coil which moves near a magnet. A potential difference is induced which makes a current flow. A loudspeaker works the opposite way. A current flowing through the coil causes it to move. The attached diaphragm moves causing vibrations in the air.

1. Explain how the vibration of a drum becomes a sound that you can hear when you hit the drum. [3]
2. Explain how the speed of the vibrations is linked to the pitch of the sound. [2]

1. *When you hit the skin of the drum, it moves backwards and forwards[1], pushing the air molecules backwards and forwards into a series of compressions and rarefactions[1] that are transmitted as a sound wave[1].*

2. *A faster vibration has a higher frequency[1], and a higher pitched note has a higher frequency, so a faster vibration will have a higher pitch[1].*

# WAVES FOR DETECTION AND EXPLORATION

Different types of wave can be used to explore and detect structures that we cannot see using differences in velocity, absorption and reflection in solids and liquids.

## Ultrasound

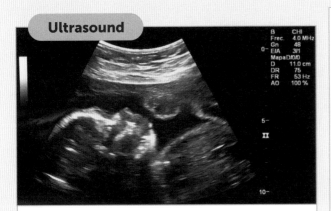

**Ultrasound waves** have a higher frequency than the upper limit of hearing for humans (over 20kHz). When ultrasound waves meet a boundary between two substances with different density they are partially reflected. The time taken for these reflections to reach a detector determines how far away the boundary between the substances is.

Ultrasound waves are used for **medical imaging**, such as observing internal organs and checking the development of a foetus during pregnancy), and in industry to find cracks and gaps in structures such as pipelines and aircraft.

**Echo sounding** uses high frequency sound waves to detect objects in deep water or to measure the depth of the water. The sounds are sent out and the time taken to receive the echo is measured.

## Seismic waves

**Seismic waves** are produced by earthquakes. There are two types of seismic wave that provide evidence for the structure and size of the Earth's core.

**P-waves** are longitudinal and travel faster in a solid than a liquid so they refract when passing from one to another.

**S-waves** are transverse and can only travel through solids, not liquids.

The detection of P-waves and S-waves after an earthquake is how we know that the centre of the Earth has a liquid outer core and solid inner core.

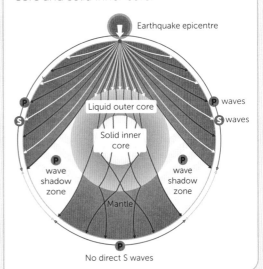

1. Give **one** advantage of using ultrasounds for medical imaging rather than gamma or X-rays. [1]
2. Suggest how ultrasound can be used for cleaning bike chains in a liquid. [2]
3. Give the equation that is used to determine the depth of water using echo sounding. [1]

   1. *Ultrasound does not expose a person to ionising radiation.[1]*
   2. *The ultrasound waves transfer vibrations through the liquid[1] which cause the dirt to loosen and fall off the chain[1].*
   3. *distance = speed × time.[1]*

# TYPES OF ELECTROMAGNETIC WAVES

Electromagnetic waves are transverse waves that transfer energy from the source of the waves to an absorber.

## Properties of electromagnetic waves

- Transverse waves
- Transfer energy from the **source** of the waves to an **absorber**.
- Form a continuous spectrum.
- Do not need a medium to travel through.
- All travel at the same velocity through a **vacuum** ($3 \times 10^8$ m/s).
- Are grouped in terms of their wavelength and their frequency.

A vacuum means there are no particles at all.

## The electromagnetic spectrum

Electromagnetic waves form a continuous spectrum.

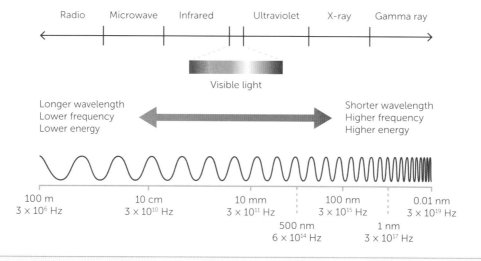

1. Which of the electromagnetic waves can be detected by the human eye? [1]
2. Which electromagnetic wave has the longest wavelength? [1]
3. Which electromagnetic wave has the highest frequency? [1]
4. Give **two** electromagnetic waves that have lower energy than visible light. [2]

   1. *Visible light.*[1]
   2. *Radio waves.*[1]
   3. *Gamma rays.*[1]
   4. *Any two from: infrared / microwave / radio.*[2]

# PROPERTIES OF ELECTROMAGNETIC WAVES 1

Waves change speed when they move from one substance into another. This can also cause a change in direction. This change in direction is called **refraction**.

## Ray diagrams

When the angle of incidence of light at a boundary is less than 90°:

- the light is refracted towards the normal when it goes from air to another material like glass or water
- the light is refracted away from the normal when it goes from a material such as glass or water to air

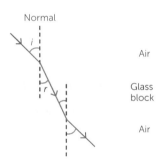

Light travels faster in air than in water. Describe what happens to light as it passes from water to air at an angle of incidence of 40°. [3]

*The light refracts and changes direction[1] and bends away from the normal[1] at an angle of refraction greater than 40° and less than 90°.[1]*

## Explaining refraction    Higher Tier

Different substances may absorb, transmit, refract or reflect electromagnetic waves in ways that vary with wavelength.

Refraction is caused by the difference in velocity of the wave in different substances. For example, light travels more slowly in air and glass than it does in water. The shorter the wavelength of an electromagnetic wave, the more it is refracted.

A **wave front** is a line that joins all the points on a wave that are moving up and down together at the same time. As the wavefronts meet the boundary between air and glass, they slow down as they enter the glass. The left-hand end of the wavefront meets the boundary and slows down before the right-hand end, and this causes the direction of the wavefront to change. The wavelength gets shorter, but the frequency remains the same.

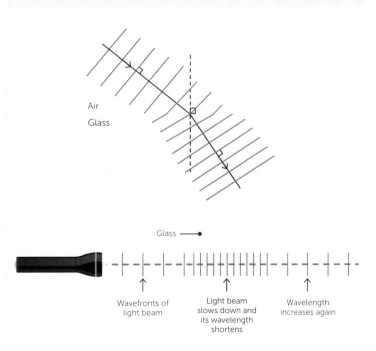

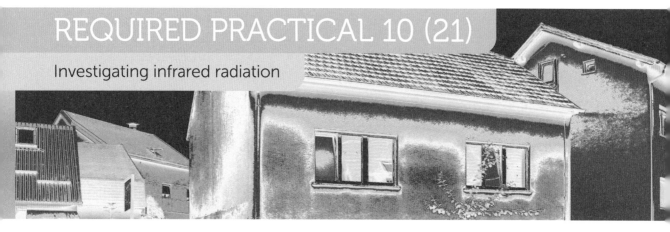

This practical activity helps you to record accurate temperatures and observe the effects of infrared radiation interacting with different surfaces.

## Emission

All objects emit infrared (IR) radiation. The hotter the surface, the more radiation is emitted. The amount of IR radiation emitted also depends on the type of surface. We cannot see IR radiation so we use an **IR detector** or observe temperature changes near the surface.

An IR detector measures the amount of infrared radiation emitted from each surface. A thermometer can also be used to measure the temperature of the air near the surface. A detector has a higher **resolution** than a thermometer so will be better for detecting small differences in the amount of radiation emitted.

A metal box called a **Leslie cube** is used for this practical. The bottom of the box is placed on a heat proof mat and the top has a hole with a lid so that very hot water can be placed inside.

The box has different surfaces on the four vertical sides, typically:

- matt black
- shiny black
- shiny silver
- matt white.

As it is a cube, all surfaces have the same surface area. The amount of infrared emitted is recorded and compared for each surface. The higher the value the greater the emission.

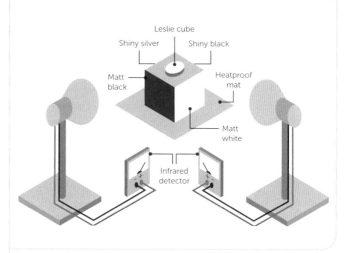

## Absorption

We can also use a Leslie cube to investigate how well the different surfaces absorb IR radiation.

Fill the Leslie cube with a known volume of cold water. Put a thermometer in the top of the cube and plug the hole with a bung.

Place a radiant heater a known distance (about 10 cm) from one surface. Switch the heater on and record the temperature of the water every 30 seconds for about 10 minutes.

Repeat for the other three faces, replacing the water each time with the same volume of cold water and making sure that the radiant heater is the same distance from the face.

| | | Dull black | Shiny black | Matt white | Shiny silver | |
|---|---|---|---|---|---|---|
| Emission | Good | | | | | Poor |
| Absorption | Good | | | | | Poor |
| Reflection | Poor | | | | | Good |

1. (a) Give **two** control variables for the absorption investigation. [2]

   (b) Explain why you need to control the variables you have named. [2]

2. A student uses a Leslie cube to observe IR emission.

   (a) The Leslie cube is left for a short while after the hot water is added before recording the emission. Explain why. [2]

   (b) Suggest the best type of graph to present the results. Explain your choice. [3]

   (c) Describe how standard laboratory thermometers can be used to investigate the emission from the Leslie cube. [4]

1. (a) *Any two from: distance of the IR source from the surface / intensity of IR source / volume of water used each time.[1]*

   (b) *To keep them constant[1] so that it is a fair test[1]*

2. (a) *It allows the surfaces to heat up to the same temperature as the water[1] so that the surfaces all have the same starting temperature.[1]*

   (b) *A bar chart[1] because the independent variable is a categoric variable[1] so a line graph cannot be plotted.[1]*

   (c) *Paint four[✓] thermometer bulbs matt black.[✓] Place a thermometer near each surface[✓] and at the same distance[✓] from each surface. Make the distance small[✓]to increase the IR radiation incident on the thermometer bulb.[✓] Record the temperature rise for each surface[✓] over a set time[✓]. The thermometer reading that reaches the highest temperature[✓] indicates the surface with the greatest emission[✓].*

   *This question should be marked in accordance with the levels based mark scheme on page 170.*

# PROPERTIES OF ELECTROMAGNETIC WAVES 2

## Changes in atoms

When electromagnetic waves are **generated** or **absorbed**, by atoms the **energy levels**, or **nucleus**, of an atom changes. This happens over a wide frequency range.

- When an electromagnetic wave is absorbed, it can cause an electron to move to a higher level.
- When an electron returns to its original energy level it can generate an electromagnetic wave.
- Gamma rays are generated by a change in the nucleus of a radioactive atom.

## Hazardous effects of UV, X-rays and gamma rays

The hazardous effect that some electromagnetic waves can have on human body tissue depends on both the **type** of radiation and the size of the **dose**.

Radiation dose is a measure of the risk of harm resulting from exposure of the body to a source of radiation. It is measured in sieverts (Sv).

1000 millisieverts (mSv) = 1 sievert (Sv)

Ultraviolet waves, X-rays and gamma rays have higher energy, so exposure carries more risk. **Ultraviolet waves** can cause skin to age prematurely and increase the risk of skin cancer. The use of sunscreen to block UV waves is recommended, and tanning beds need to be used with caution. **X-rays** and **gamma rays** are both **ionising** radiation. This means they remove electrons from atoms. This can cause mutations in cells and genes, increasing the risk of cancer.

## Radio waves        Higher Tier

Radio waves can be produced by **oscillations of electrons** in electrical circuits. Radio waves can be absorbed by a conductor such as an aerial. This can induce the electrons in the conductor to oscillate and generate an alternating current with the same frequency as the absorbed radio waves. This is how terrestrial TV and radio is broadcast and received.

1. Complete the sentences.                                                                                          [2]

    An electron can move to a higher energy level when electromagnetic radiation is _____.

    When an electron moves to a lower energy level electromagnetic radiation can be _____.

2. The radiation dose from a single X-ray is about 0.1 mSv. A fatal dose of radiation is about 5 Sv. Explain why hospital staff working with X-rays need to be shielded when a patient is having an X-ray.    [2]

    1.  *absorbed[1] emitted[1]*

    2.  *As hospital staff could see many patients each day, this could lead to a high dose of radiation over a period of time.[1] So staff shield to minimise the dose of X-rays they are exposed to, to reduce the potential damage that could be caused.[1]*

# USES AND APPLICATIONS OF ELECTROMAGNETIC WAVES

## The electromagnetic spectrum

Electromagnetic waves have very many applications. Some examples are shown in the table.

| Wave | Uses | **Higher Tier only** Suitability of each electromagnetic wave |
|---|---|---|
| Radio waves | Television and radio, communication with ships, aeroplanes and satellites | • Travel long distances before being absorbed. • Reflect off the ionosphere in the atmosphere so can be sent very long distances around the earth. |
| Microwaves | Satellite communications, mobile phones, cooking food | • Pass through the Earth's atmosphere without reflection or refraction. • Food contains a lot of water which absorbs microwaves and heats up. |
| Infrared | Electrical heaters, cooking food, infrared cameras, TV remote controls | • Emitted by hot objects and easily absorbed by surfaces. • Check for heat being emitted from animals or heat loss from buildings. |
| Visible light | Fibre optic communications | • Thin strands of glass which reflect visible light inside to carry pulses of light in cables. • Short wavelength means a lot of information can be transmitted. |
| Ultraviolet | Energy efficient lamps, sun tanning | • Short wavelength UV converted to visible light inside the bulb. • Tanning beds use UV light to reproduce the effect of the Sun on skin. |
| X-rays and gamma rays | Medical imaging, medical treatments | • X-rays and gamma rays have high energy and are very penetrating. • Both pass through tissue but X-rays are absorbed by bones, so bones show up on photographic film. • Gamma rays can also be used to target and destroy cancers. |

1. Give **two** types of electromagnetic waves used for cooking.  [1]
2. Give **two** types of electromagnetic waves used for communicating.  [1]
3. **Higher Tier only:** Describe why X-rays and not gamma rays are used to diagnose broken bones.  [2]

1. *Microwave and infrared.[1]*
2. *Any two from: radio waves, microwaves, visible light.[1]*
3. *X-rays can pass through surrounding tissue but are absorbed by bone.[1] Gamma rays pass straight through both tissue and bone.[1]*

# LENSES

A **lens** is a transparent object with curved sides, usually made from glass or plastic. As light travels through the lens, it refracts at both surfaces.

## Convex and concave lenses

There are two main types of lens, **convex** and **concave**. **Ray diagrams** are used to show the formation of an image by a lens. Symbols are used for the lenses when drawing ray diagrams.

A real image can be seen on a screen. An image is formed when light rays converge and are focused on a screen.

A virtual image cannot be seen on a screen. An image is formed when light rays appear to come from a point.

The **magnification** produced by a lens can be calculated using the equation:

$$\text{magnification} = \frac{\text{image height}}{\text{object height}}$$

Magnification is a ratio and has no units.

> You need to be able to select this equation from the equation list and apply it.

| Convex lens | Concave lens |
|---|---|

In a **convex** lens, parallel rays of light **converge** (come together) at the **principal focus**.

The distance from the lens to the principal focus is called the **focal length**.

The image produced by a convex lens can be either **real** or **virtual**.

With a **concave** lens, parallel rays of light **diverge** (spread out) as though they have come from the principal focus.

The image produced by a concave lens is always **virtual**.

> Remember to use the same units for both heights – usually millimetres or centimetres.

## Real images

Real images are always formed by a convex lens. To show how an image is formed by a lens, at least two rays are needed. Real rays are shown as solid lines.

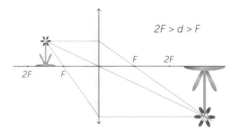

When the image is far away from the lens the real image is smaller than the object and between one and two focal lengths from the lens.

When the object is at a distance of two focal lengths (2F), the image and the object are the same size. When the object is between one and two focal lengths from the lens, the image is further than two focal lengths from the lens and is magnified.

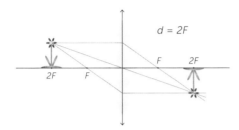

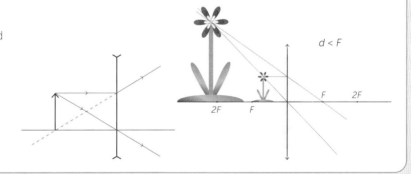

## Virtual images

Virtual images can be formed by both convex and concave lenses. Virtual rays are shown as dashed lines.

1. Draw a ray diagram to show the path of parallel light through a convex lens. [2]
2. Describe the difference between a real image and a virtual image. [1]
3. A convex lens has a magnification of 15. Calculate the height of the image it will produce of an object that is 2.4 cm high. [3]
4. Describe where you would put the object to use a convex lens as a magnifying glass. [1]

1. *Correct symbol for convex lens.*[1] *3 parallel rays converging to a point and passing through.*[1]
2. *A real image can be projected onto a screen and a virtual image cannot be as it is not actually there.*[1]
3. *magnification = image height / object height*
   *15 = image height / 2.4*[1]   *Image height = 15 × 2.4*[1] *= 36 cm*[1]
4. *Between the lens and its principal focus.*[1]

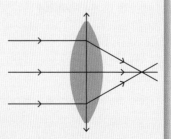

# VISIBLE LIGHT

The visible light spectrum is made up of a continuous range of different colours of light. Each colour has its own narrow band of wavelength and frequency. The colours always appear in the same order: **Red orange yellow green blue indigo violet**.

The colour of objects depends on whether the different wavelengths are reflected, transmitted or absorbed at a surface.

## Reflection and absorption

Visible light can be reflected in two ways:

- **Specular reflection** is when the light is reflected from a smooth surface in a single direction.
- **Diffuse reflection** is when the light is reflected from a rough surface which causes the light to scatter. The incident ray is reflected at many different angles

Specular Reflection

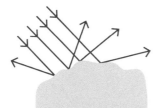

Diffuse Reflection

The **colour of an opaque object** is determined by which wavelengths of light are more strongly reflected.

If all wavelengths are reflected equally the object appears **white**.

If all wavelengths are absorbed the objects appears **black**.

If the red wavelength is reflected and all others are absorbed then the object appears **red**.

A coloured object will appear black if no light of the same colour can be reflected from it. Wavelengths that are not reflected are absorbed.

White surface

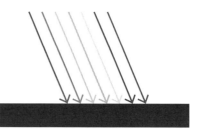

Black surface

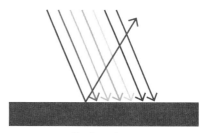

Red surface

Red filter

## Transmission

When an object transmits light, the wave continues through the material. **Transparent** materials such as air, glass and water are particularly good at transmitting light waves. **Translucent** materials such as tissue paper and frosted glass transmit some light and absorb or reflect the rest. **Opaque** objects transmit no light so absorb or reflect all the light.

When white light passes through a **colour filter** all the wavelengths are absorbed apart from the wavelength with the same colour as the filter. This colour is transmitted through the filter.

1. Describe what happens when white light is incident on a green filter. [1]
2. When will an object appear white? [1]

   1.The green light will be transmitted, and all the other colours will be absorbed.[1]

   2. When all of the wavelengths in visible light are equally reflected.[1]

# BLACK BODY RADIATION

## Emission and absorption of infrared radiation

> Incident means radiation that reaches the body.

**All** bodies (or objects) emit and absorb infrared radiation whatever their temperature. However, the hotter a body is, the more infrared radiation it emits in a given time.

A **black body** does not reflect or transmit any radiation. A perfect black body is an object that absorbs **all** of the radiation incident on it. No radiation is reflected or transmitted at all. A perfect black body is also the best possible emitter, as a good absorber is also a good emitter.

## Perfect black bodies and radiation

- All bodies emit radiation.
- The **intensity** and **wavelength** of any emissions depends on the temperature of the body.
- Hotter bodies emit shorter wavelength radiation than cooler ones.

The graph shows how the temperature of an object affects the wavelength and intensity of the radiation it emits.

As the temperature of an object increases it emits more radiation with shorter wavelengths.

This is why very hot objects emit visible light and can become red or white hot for example.

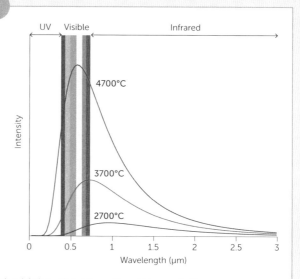

The graph also shows that the hotter the object, the higher the intensity of the radiation.

## The temperature of a body    Higher Tier

The **temperature** of a body depends on the relative rates that radiation is being emitted and absorbed by the body.
- A body will stay at a **constant temperature** when it absorbs and emits radiation at the **same rate**.
- A body gets **hotter** when it **absorbs** radiation **faster** than it emits radiation.
- A body **cools** down when it **emits** radiation **faster** than it absorbs radiation.

The balance between incoming radiation absorbed and radiation emitted can be shown by considering these common hot and cold bodies.

The temperature of a hot drink is higher than the surroundings so it emits more infrared radiation than it absorbs. The hot drink will cool down until it reaches the same temperature as the surroundings and the rate of emission and absorption are the same.

The temperature of a cold ice lolly is lower than the surroundings so it absorbs more infrared radiation than it emits. The temperature of the lolly rises until it is the same as the surroundings.

The Earth emits only infrared radiation because the temperature of the Earth is much lower than the temperature of the Sun.

The effective temperature of the Sun is about 5500°C, so it emits ultraviolet, visible light and short wavelength infrared radiation. Some of this radiation is reflected away by clouds and the atmosphere so it does not reach the surface of the Earth. The rest is **absorbed** by the surface of the Earth. This increases the internal energy of the Earth and the surface temperature increases.

The Earth then **emits low frequency IR** radiation back into space. Some of this radiation is absorbed by greenhouse gases (e.g. methane and carbon dioxide) in the atmosphere. The greenhouse gases emit infrared radiation, and some of this radiation is reabsorbed by the Earth.

The temperature of the Earth is determined by the overall balance of the absorption and emission of radiation. The temperature of the Earth will remain constant if it absorbs the same amount of radiation as it emits.

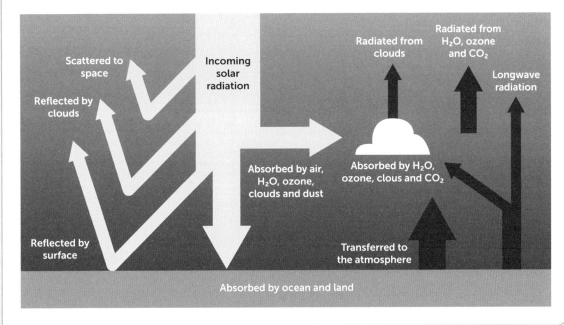

1. What is the relationship between the temperature and the infrared radiation emitted by a body?  [1]

2. **Higher Tier only:** Carbon dioxide in the atmosphere absorbs infrared radiation that is emitted from the surface of the Earth that would otherwise be emitted into space. Explain how increasing levels of carbon dioxide in the atmosphere is changing the temperature of the Earth.  [6]

1.  *As the temperature of the body increases, the rate of IR radiation it emits also increases.*[1]

2.  *The temperature of a body is determined by the overall balance between the rates of absorption and emission of radiation.*[M] *The Earth absorbs ultraviolet, visible light and infrared radiation from the Sun*[M] *Some of this radiation is reflected.*[M] *The rest is absorbed and causes the temperature of the surface of the Earth to increase.*[M] *All bodies emit radiation*[M] *so radiation is emitted from the Earth back into space*[M] *as infrared radiation*[M]. *As the levels of carbon dioxide increase, more of this infrared radiation is absorbed in the atmosphere than being emitted into space.*[M] *This means that the Earth is absorbing more radiation than it is emitting*[M] *so the temperature of the Earth is increasing*[M]. *Note: This question should be marked in accordance with the levels based mark scheme on page 170.*

# EXAMINATION PRACTICE

01 This question is about the properties of waves.

01.1 The diagram represents a longitudinal wave. Label a **wavelength**, a **compression** and a **rarefaction**. [3]

01.2 The period of a wave is 0.4 s. Calculate the frequency of the wave. Include the unit. [4]

01.3 A wave with a frequency of 160 Hz travels at a speed of 8.0 m/s.
Calculate the wavelength of the wave. [3]

01.4 Describe a method for two students to measure the speed of sound in air using echoes.
Include steps to reduce the effect of errors. [6]

02 This question is about measuring properties of waves in liquids and solids.

02.1 A ripple tank has a total length of 0.56 m. 28 waves are counted on the ripple tank.
Calculate the wavelength of the waves. [2]

The table shows the frequency of a wave in a ripple tank observed by four students A, B, C and D.

| Student | Frequency in Hz |
|---------|-----------------|
| A | 2.9 |
| B | 3.3 |
| C | 3.2 |
| D | X |
| Mean | 3.2 |

02.2 Calculate value X. [2]

02.3 The wave has a wavelength of 0.075 m. Calculate the speed of the wave. [2]

02.4 Explain why taking a video of the waves on a ripple tank is better than taking a photograph of the waves. [3]

The diagram shows a standing wave generated on a string. The wave has a frequency of 12.0 Hz.

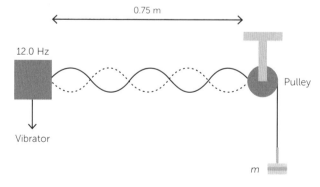

02.5 Calculate the speed of the wave in the string. [4]

03    **Physics only:** This question is about reflection and refraction.

03.1    Describe what can happen to a wave at the boundary between two materials.    [2]

03.2    Complete the ray diagram to show the path of a light ray that is reflected from the mirror. The angle of incidence is 40°.    [2]

A student investigates the path of light from a single slit ray box.
The diagrams show the path when the light ray is incident on two blocks made of different transparent materials.

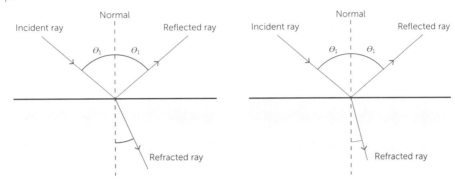

03.3    Measure the angles and complete the table.    [2]

|  | Material 1 | Material 2 |
|---|---|---|
| Angle of incidence |  | 2.9 |
| Angle of reflection |  | 3.3 |
| Angle of refraction |  | 3.2 |

03.4    Write **two** conclusions for the investigation.    [2]

04    This question is about electromagnetic waves.

04.1    Give **two** features of all electromagnetic waves.    [2]

04.2    Complete the ray diagram to show what happens as light travels from air to water. Include labels.    [2]

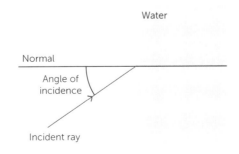

04.3 **Higher Tier only:** The diagram shows wavefronts travelling from deep water to shallow water.
Explain what happens to the wavefronts as they travel from deep water to shallow water. [3]

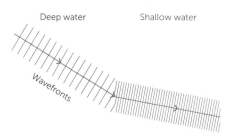

A student uses a Leslie cube to investigate the emission of infrared radiation from different types of surface.

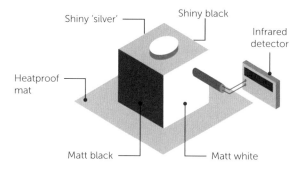

04.4 Give the independent and dependent variables for this investigation. [2]

04.5 Give **one** control variable. [1]

04.6 The bar chart shows the results. Draw **one** line from each bar label to the type of surface. [2]

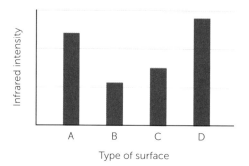

| Bar label |
| --- |
| A |
| B |
| C |
| D |

| Type of surface |
| --- |
| Matt black |
| Matt white |
| Shiny black |
| Shiny silver |

04.7 Give the type of electromagnetic wave that is the most damaging to human tissue. [1]

04.8 **Higher Tier only:** Describe how radio waves can be generated. [3]

04.9 Give **two** uses of infrared radiation. [2]

04.10 **Higher Tier only:** Explain why radio waves are used for transmitting terrestrial television signals. [2]

05 **Physics Higher Tier only:** This question is about sound waves and seismic waves.

05.1 What is the range of normal human hearing? Tick **one** box. [1]

☐ 2 Hz to 2000 Hz

☐ 20 Hz to 20000 Hz

☐ 200 Hz to 200 000 Hz

☐ 20 kHz to 200 kHz

05.2 Explain why sound travels faster in a solid than a gas. [2]

05.3 Compare the properties of P-waves and S-waves. [2]

05.4 High frequency sound waves are used to determine the depth of the sea.
Calculate the depth of the water when a reflected ultrasound pulse is detected 2.5 s after it was sent.
The speed of sound in water in 1500 m/s. [3]

06 **Physics only:** This question is about visible light.

06.1 Give **two** differences between the effect of a convex and a concave lens on light. [2]

06.2 Calculate the magnification of a lens that produces an image that is 2.6 cm high when the object is 0.052 mm high. [3]

06.3 Draw a ray diagram to show the path of parallel light through a concave lens. [2]

06.4 Give the colour a blue object will look in each colour of light:

Blue light: _____ [1]

Green light: _____ [1]

White light: _____ [1]

07 **Physics only:** This question is about black body radiation.

07.1 Define a perfect black body. [2]

07.2 Describe how the temperature of an object affects the wavelength **and** the intensity of the radiation it emits. [2]

07.3 **Physics Higher Tier only:** What happens to the temperature of an object that emits radiation at a greater rate than it absorbs radiation? Tick **one** box. [1]

☐ It decreases

☐ It remains constant

☐ It increases

# POLES OF A MAGNET

Magnets are made from a magnetic material and are either permanent or induced:

- A **permanent** magnet produces its own magnetic field.
- An **induced** magnet is a temporary magnet created when a magnetic material becomes a magnet while in a magnetic field.

## Permanent magnets

Bar and horseshoe magnets are examples of permanent magnets.

The ends of a magnet are called the **poles** and are described as **north (N)** or **south (S)**. The poles are the places where the magnetic forces are strongest.

A permanent magnet will always exert a force on another magnet. The type of force exerted depends on the interacting poles. These are non-contact forces as the magnets do not need to touch to experience the force.

When two magnets are brought close together, they exert a force of attraction or repulsion on each other.

- two **like** poles will **repel** each other (N/N; S/S).
- two **unlike** poles will **attract** each other (N/S).

Permanent magnets also attract magnetic materials.

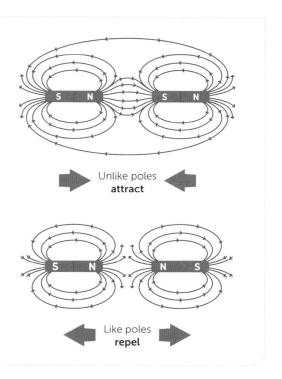

Unlike poles **attract**

Like poles **repel**

## Induced magnets

When a magnetic material (that is not a permanent magnet) is placed in a magnetic field, a magnetic field is **induced**, and the material becomes a **temporary** magnet. When the material is removed from the magnetic field the induced magnet quickly loses nearly all of its magnetism.

An induced magnet will always cause a force of attraction.

1. Magnetic materials are affected by magnets and attracted to both poles of a magnet. Which material is **not** magnetic? Tick **one** box. [1]

   ☐ Cobalt   ☐ Copper   ☐ Iron   ☐ Nickel   ☐ Steel

2. How can it be shown that a magnetic material is a permanent magnet? [1]

3. Describe the forces on Magnet 2. [2]

   | S | N |   | S | N |   | N | S |
   Magnet 1      Magnet 2      Magnet 3

1. *Copper[1]*

2. *It will attract and also repel another magnet.[1]*

3. *The forces are non-contact forces[1] Magnet 2 experiences a force of attraction from Magnet 1[1] and a force of repulsion from Magnet 3.[1]*

# MAGNETIC FIELDS

A magnetic field is the region around a magnet where a non-contact magnetic force acts on either:
- another magnet
- or a magnetic material (iron, steel, cobalt and nickel)

## Strength and direction of the magnetic field

The **strength** of the field depends on the **distance** from the magnet. The field is strongest at the **poles** of the magnet.

The **direction** of the magnetic field at any point is given by the direction of the force that would act on another north pole placed at that point. The direction of a magnetic field line is from the north (seeking) pole of a magnet to the south (seeking) pole of the magnet.

Remember the force between two magnets can be attractive or repulsive. The force between a magnet and a magnetic material is always attractive.

## Magnetic field pattern of a bar magnet

A magnetic **compass** contains a small bar magnet. The north needle of a compass is attracted to the south pole of a magnet. The magnetic field lines of a magnet can be plotted using a compass.

When a compass is away from other magnets it always points towards the same place on Earth. This shows that the **Earth** must have a **magnetic field**. The compass needle points in the direction of the Earth's magnetic field and provides evidence for a **magnetic core** in the Earth.

These six steps describe how to plot the magnetic field pattern of a bar magnet on a piece of paper using a plotting compass. Put the six steps in the correct order.                    [3]

A  Repeat this until the south pole of the magnet is reached;

B  Join the crosses to show the field lines;

C  Move the compass so the south pole of the needle touches this cross;

D  Place the compass near the N pole of the magnet;

E  Mark a small cross where the compass needle is pointing;

F  Repeat for different positions from the N pole of the magnet.

*The correct order is: D, E, C, A, B, F. 2 marks for 5 in correct order, 1 mark for 4 in correct order.*

Notice how the strength and direction of the field changes from one point to another.

Check where the north pole is, the magnet could be either way around.

# ELECTROMAGNETISM

**Electromagnetism** is the effect where a magnetic field is produced by result of an electric current flowing through a wire. It is a fairly simple effect yet has many useful applications which affect the way we live.

## Current in a wire

A **magnetic field** is produced around a conducting wire when a **current** flows through the wire. The **strength** of the magnetic field depends on both the **current** and the **distance** from the wire.

- As the current in the wire increases, the strength of the magnetic field increases: they are directly proportional.

- As the distance from the wire increases, the strength of the magnetic field decreases: they are inversely proportional.

The **direction** of the magnetic field depends on the **direction** of the current in the wire.

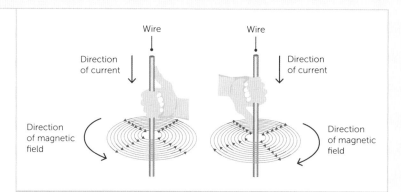

Right hand grip rule for a wire:
- thumb points in the direction of the current
- fingers point in the direction of the field

Conventional current direction is from the positive terminal to the negative terminal of the supply.

## Solenoid

Making a coil of wire increases the strength of the magnetic field produced by a current through the wire. A **coil** of wire used in electromagnetism is called a **solenoid**.

- The magnetic field **inside** a solenoid is **strong** and **uniform**.

- The magnetic field **around** a solenoid has a similar shape to the magnetic field around a **bar magnet**

To increase the **strength** of the magnetic field of a solenoid:

- increase the **current** in the wire

- increase the **number of turns** of wire in the coil

- add an **iron core** inside the coil.

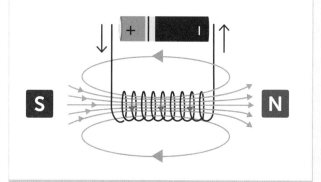

Right hand grip rule for a solenoid:
- fingers curl in the direction of the current
- thumb points in the direction of the north pole

## Electromagnet

An **electromagnet** is a **solenoid** with an **iron core** inside the coil.

Electromagnets have advantages over permanent magnets:

- they can be switched on and off which is useful for moving magnetic items.
- their strength can be altered by varying the current.

Electromagnets have very many useful applications.

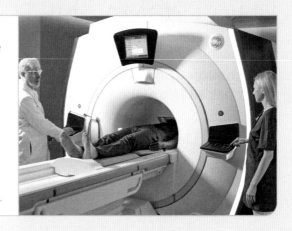

1. Describe how the magnetic effect of a current can be demonstrated. [2]

2. Give **three** ways to increase the strength of a magnetic field around a solenoid. [3]

3. (a) A student investigates the effect of current on the strength of an electromagnet as shown.

    Give the independent, dependent and control variables. [3]

    (b) Describe how the student could carry out this experiment using the equipment shown in the diagram. [6]

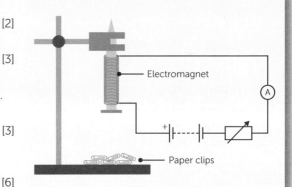

1. *Plotting compasses or iron filings can be placed near a current carrying wire.[1] The effect on the direction of the compass needle/pattern made by the iron filings gives evidence for a magnetic field.[1]*

2. *Increase the current in the wire[1] increase the number of turns in the solenoid[1], or add an iron core[1].*

3. (a) *Independent variable: current in the coil[1]; dependent variable: number of paper clips picked up[1]; control variables: number of coils of wire; size/material of paper clips; distance the electromagnet is held from the paper clips[1].*

    (b) *Adjust the power supply / variable resistor to choose a suitable value for the current e.g. 0.1A[M] Hang as many paper clips as possible from the electromagnet.[M] Record the number of paperclips in a table with the current.[M] Turn off the electromagnet.[M] Repeat the test at this current[M]. Adjust the power supply/variable resistor to increase the current to e.g. 0.2A.[M] Repeat the method to hang paper clips.[M] Collect readings for at least five different values of current.[M] Calculate a mean.[M] Check for any anomalous results.[M] The more paper clips the electromagnet picks up, the stronger the electromagnet is.[M]*

    *This question should be marked in accordance with the levels based mark scheme on page 170.*

# ELECTROMAGNETIC DEVICES

Diagrams and circuits with an electromagnet can be interpreted to explain how the electromagnet is used. The same principles can be applied to other examples which make use of the ability to switch magnetic attraction on and off.

## How an electromagnetic device works

For all the devices, when the electromagnet is switched on, the current produces a magnetic field. This is used to attract a piece of magnetic metal which is free to move on a spring or a pivot.

In a **relay**, this closes a switch to complete another circuit.

In a **door lock**, this pulls a bolt out of a door so it unlocks.

In an **electric bell** this pulls a clapper against the bell. As the clapper moves it breaks the circuit. This turns off the magnet and the clapper is released and reconnects the circuit. This continues and the magnet is switched on and off so the bell rings until the circuit is switched off with the switch.

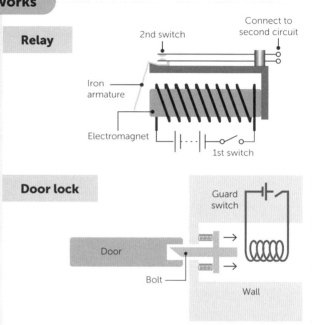

Relay

Door lock

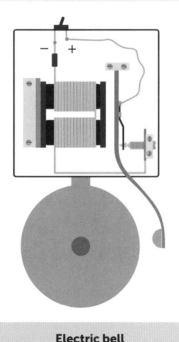

**Electric bell**

Explain how the ignition switch in the diagram is used to switch on the starter motor.    [3]

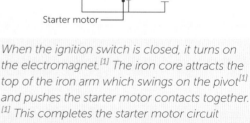

*When the ignition switch is closed, it turns on the electromagnet.[1] The iron core attracts the top of the iron arm which swings on the pivot[1] and pushes the starter motor contacts together. [1] This completes the starter motor circuit allowing a current to flow so the starter motor works.[1]*

# FLEMING'S LEFT-HAND RULE

A conductor produces a magnetic field when a current flows through it. When the conductor is placed at **right angles** to another magnetic field the two magnetic fields interact. The conductor exerts a force on the magnet and the magnet exerts a force on the conductor, and this is known as the **motor effect**.

## Calculating the size of the force on a conductor

For a conductor carrying a current at right angles to a magnetic field:

$$\text{force} = \text{magnetic flux density} \times \text{current} \times \text{length}$$

$F = BIl$

$F$ = force in newtons, N

$B$ = magnetic flux density in tesla, T

$I$ = current in amperes (amps), A

$l$ = length of wire in metres, m

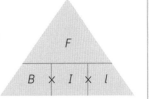

You need to be able to select this equation from the equation list and apply it.

A conductor usually means a metal wire.
Magnetic flux density is a measure of the strength of the magnetic field.

## Fleming's left-hand rule

You can use Fleming's left-hand rule to determine the direction of the force on a conductor in a magnetic field. This is a handy way to represent the relative orientations of the force or movement (**t**humb), the magnetic field (**f**irst finger) and the **c**urrent in the conductor (**s**econd finger).

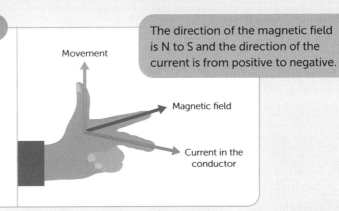

The direction of the magnetic field is N to S and the direction of the current is from positive to negative.

1. Calculate the magnetic flux density of a field that exerts a force of 0.2 N on a wire that is 1.0 m long and carrying a current of 0.8 A. [3]

2. Explain what will happen when a wire carrying a current is placed south to north in a magnetic field. [2]

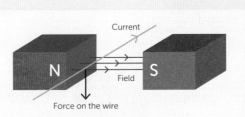

1. $F = BIl$    $0.2 = B \times 0.8 \times 1.0$[1]
   $B = 0.2 / (0.8 \times 1.0)$[1] $= 0.25$ T[1]

2. Nothing[1] because the current is parallel to the magnetic field and so they will not interact as they are not at right angles to each other[1].

# ELECTRIC MOTORS

An **electric motor** is an application of the **motor effect**. When a coil of wire carrying a current is placed at right angles to another magnetic field, the force experienced by the coil causes the coil to rotate. Electric motors have multiple uses and it is very useful for an appliance to produce a turning force.

## A simple electric motor

- When the current flows in the direction shown in the diagram through the loop of wire in the magnetic field, the force acts upwards on the left side and downwards on the right side of the loop.
- This produces a clockwise moment. The left side of the loop moves up and the right side moves down.
- When the loop is vertical, the force on the loop is zero and there is no moment. At this point there is no connection to the battery because of the gap in the split-ring commutator.
- The loop has momentum and keeps moving. The split-ring commutator swaps the connections with the battery so the current starts flowing again, but now in the opposite direction through the loop. This means that the force on the loop is still upwards on the left and downwards on the right so the loop keeps moving in the same direction.
- Note that the current still flows from positive to negative. The loop has switched sides so the current flows the opposite way through the wire.

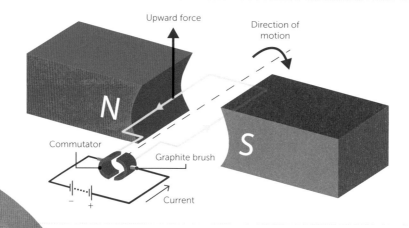

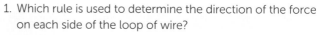

1. Which rule is used to determine the direction of the force on each side of the loop of wire?  [1]
2. Explain the purpose of a split ring commutator in an electric motor.  [2]

   1. *Using Fleming's left-hand rule.[1]*
   2. *The split-ring commutator keeps the motor rotating in the same direction[1] by allowing the current in the coil to change direction each half-turn as the coil rotates[1].*

# LOUDSPEAKERS

**Loudspeakers** also use the **motor effect**. They convert the variations of a changing **current** in an electrical circuit to the variations of air pressure that are needed to produce **sound waves**.

## Moving-coil loudspeakers and headphones

This is a simple diagram of a **moving-coil** loudspeaker found in speaker systems.

Headphones work in a similar way on a smaller scale.

A varying current in the coil produces a magnetic field in the coil. The field in the coil interacts with the magnetic field from the permanent magnet.

This produces a resultant force which moves the **cone**. This is the motor effect.

As the current varies, the force on the speaker cone varies. This causes the speaker cone to vibrate.

The vibration causes air molecules to vibrate. The movement of the air molecules produces the pressure variations needed to generate the compressions and rarefactions that cause sound waves.

Different frequencies of the varying current change the frequency that the cone vibrates. This changes the pitch of the sound.

Increasing the current increases the size of the force and also the amplitude of the vibrations.

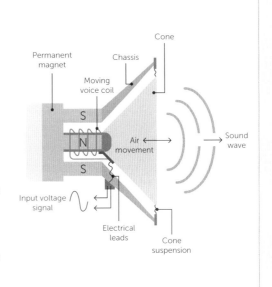

1. Explain what the motor effect is. [2]
2. Explain how the sound volume is increased in moving-coil loudspeakers and headphones. [2]

   1. *The motor effect is when a conductor carrying a current is placed in a magnetic field[1] and the interaction of the magnetic field from the magnetic and the conductor, produces a force on the conductor[1].*
   2. *Increasing the current[1] increases the amplitude of the force in the cone, so the vibrations are bigger and the sound is louder.[1]*

# INDUCED POTENTIAL

To induce means to produce.

## The generator effect

A potential difference can be **induced** across the ends of an electrical conductor such as a wire. This happens when:

- the moving conductor cuts through a magnetic field, **moving relative** to the field.
- a moving magnet produces a **changing magnetic field** around a stationary conductor.

When the conductor is part of a complete circuit such as a loop of wire, a **current** is induced in the conductor. This is called the **generator effect**.

An induced current generates a magnetic field that opposes the original change, either the movement of the conductor or the change in magnetic field.

There will only be an induced potential or current when there is relative movement, not when the conductor is stationary, and not when the conductor moves along rather than through the field lines.

When the conductor moves in the opposite direction, or the direction of the magnetic field is reversed, the **direction** of the induced potential or current also reverses. This changing direction generates an alternating potential or current.

The **size** of the induced potential or current in a wire can be increased by:

- increasing the **strength** of the magnetic field causing the induction in the wire

- increasing the **speed** the wire moves through the magnetic field

- increasing the **number of coils** the wire is shaped into.

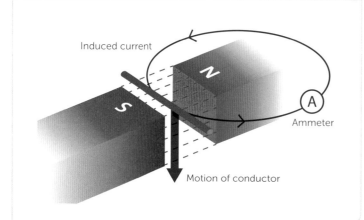

Induced current

Ammeter

Motion of conductor

1. Explain the effect shown in the diagrams. [5]

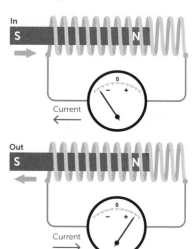

In
S    N
Current

Out
S    N
Current

2. Explain why inducing a current makes it harder to move the magnet. [2]

*1. Moving the magnet in and out of the coil produces a changing magnetic field[1] which induces a current in the coil as it is part of a complete circuit[1]. As the N pole enters the coil the induced current flows in one direction.[1] As the magnet is pulled out it reverses the direction of the relative movement of the magnetic field and the coil[1] which also reverses the direction of the current[1].*

*2. The induced current also generates its own magnetic field.[1] This field opposes the movement of the magnet that is causing the induced current.[1]*

# MICROPHONES

**Microphones** use the **generator effect**. Microphones convert the variations of air pressure in **sound waves** into variations in **current** in electrical circuits.

## Moving coil microphones

A microphone is covered by a flexible piece of material called a **diaphragm**.

The diaphragm vibrates when sound waves are incident on it.

As the diaphragm is connected to the coil, the coil also vibrates. It moves backwards and forwards in the magnetic field of the permanent magnet.

The generator effect causes a potential difference and an alternating current to be induced.

The frequency of the induced pd is the same as the frequency of the sound waves that are incident on the diaphragm.

The electrical signal produced can be passed into an amplifier to increase the volume of the sound. It can also be passed into a moving-coil loudspeaker or headphones, or into a recording device.

Notice the connection between microphones and loudspeakers. A microphone works like a loudspeaker in reverse.

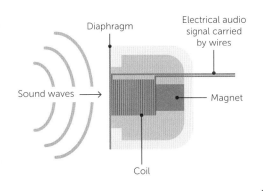

1. What effect is used in a moving-coil microphone? [1]
2. Explain why it is possible that a loudspeaker could also be used as a microphone. [2]

   1. *The generator effect.[1]*
   2. *A loudspeaker and a microphone both contain the same parts.[1] A microphone converts sound waves to an electrical signal and a loudspeaker converts and electrical signal to sound waves.[1]*

# USES OF THE GENERATOR EFFECT

The generator effect is also used to generate:
- an alternating current (**ac**) in an **alternator**
- a direct current (**dc**) in a **dynamo**

> Inducing a pd also induces a current when the conductor is a loop.

## A simple ac alternator

An alternator is a coil of wire rotating in a magnetic field so it induces a pd. Each end of the coil is connected to a metal ring called a **slip ring commutator**. This allows the current to pass out of the coil to a connected circuit.

The direction of the pd will depend on whether that side of the coil is moving up or moving down in the field as it spins. This means an **alternating current** is generated as the direction changes in one full rotation.

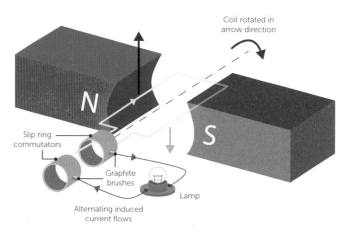

Coil rotated in arrow direction

Slip ring commutators

Graphite brushes

Lamp

Alternating induced current flows

## Graph of pd against time

The maximum pd is induced when the coil is horizontal and passing directly through the field lines at maximum speed.

When the coil is vertical it is parallel to the field so no pd is induced.

The pd is negative when the two sides of the coil move in the opposite direction to when the pd is positive.

During one full rotation, each side of the coil passes through the field twice, once up and once down.

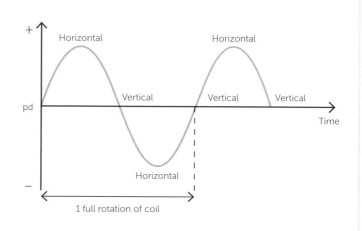

## A simple dc dynamo

A dynamo is a coil of wire rotating in a magnetic field so it induces a pd. Each end of the coil is attached to one **split ring commutator**.

As the coil spins, the wires swap sides so each side of the commutator always has the end of the coil touching it as it goes up or down.

This means the direction of the pd does not change and a **direct current** is generated.

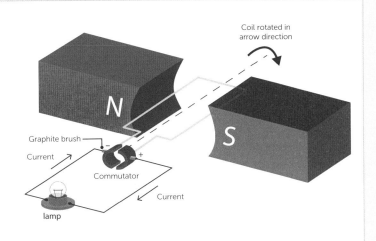

## Graph of pd against time

The maximum pd is induced when the coil is horizontal and passing directly through the field lines at maximum speed.

When the coil is vertical, it is parallel to the field so no pd is induced.

The pd is always in the same direction because the direction of the pd does not change.

During one full rotation, each side of the coil passes through the field twice, once up and once down.

The graph could be the reverse of this with the pd negative. It depends on the direction of the direct current.

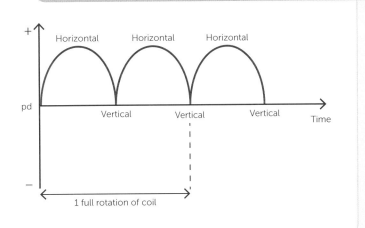

1. Which device is used to:
   (a) produce the mains supply? [1]
   (b) light a lamp on a bicycle? [1]
2. Increasing the speed the coil rotates increases the size of the alternating current generated in an alternator.
   Give **two** other ways to increase the size of the alternating current. [2]
3. What also increases when the rotation speed of the coil increases? [1]

   1. (a) alternator[1]    (b) dynamo[1]
   2. Increase the strength of the magnetic field[1] increase the number of turns/ increase the area of the coil[1].
   3. The frequency of the alternating current also increases.[1]

# TRANSFORMERS

Transformers are used to change the potential difference of an electricity supply. They are commonly used throughout the National Grid and in devices where the pd needs to be increased or decreased.

## Types of transformer

A basic **transformer** consists of a **primary coil (p)** shown on the left, and a **secondary coil (s)** shown on the right.

Both coils are made of a conductor wound on an iron core. Each coil is in a completely separate circuit.

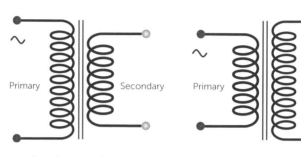

Primary                    Secondary    Primary                    Secondary

Step-down transformer              Step-up transformer

There are two main types of transformer:

- **step-up** when the primary coil has less turns than the secondary coil; the potential difference across the secondary coil is greater than the potential difference across the primary coil

- **step-down** when the primary coil has more turns than the secondary coil; the potential difference across the secondary coil is less than the potential difference across the primary coil.

## How a transformer works

An **alternating current** in the **primary coil** generates a changing magnetic field. This **changing magnetic field** is transmitted along the iron core which also increases the strength of the changing magnetic field. Iron is used as it is easily magnetised.

As the changing field passes through the **secondary coil** it induces a potential difference and an **alternating current** in this coil.

## Ratio of the potential differences

The ratio of the potential differences across the primary coil ($V_p$) and the secondary coil ($V_s$) of a transformer is equal to the ratio of the number of turns on the primary coil ($n_p$) and the number of turns on the secondary coil ($n_s$).

The potential difference is measured in volts, V and the number of turns have no units. This assumes that transformers are 100% efficient.

$$\frac{V_p}{V_s} = \frac{n_p}{n_s}$$

You need to be able to select this equation from the equation sheet and apply it.

There is always some energy loss to the surroundings so 100% efficiency is not possible. Assume it is for calculations.

Assuming a transformer is 100% efficient:

**electrical power output (secondary coil) = electrical power input (primary coil)**

Electrical power in watts (W) = $V \times I$. So:

$$V_s \times I_s = V_p \times I_p$$

where $V_s \times I_s$ is the power output (secondary coil) and $V_p \times I_p$ is the power input (primary coil).

Transformers are used to reduce energy losses in the National Grid. When the potential difference is increased, the current decreases. So large powers are transmitted at a high potential difference to reduce energy losses. If the resistance of a wire is $R$, the energy loss in the wire is given by $P = I^2R$ as $P = IV$ and $V = IR$. Doubling the current means that the energy loss in the wire increases by a factor of 4.

So the power is transmitted at high voltages and stepped up to 400 000V and carried in high voltage cables before stepping down to safer levels of 230 V for domestic and business use.

1. What effect is used in transformers? [1]
2. Which row describes the potential difference in transformers?
   Tick **one** box. [1]

|   | Step-up | Step-down |
|---|---------|-----------|
| ☐ A | $V_s > V_p$ | $V_s > V_p$ |
| ☐ B | $V_s > V_p$ | $V_s < V_p$ |
| ☐ C | $V_s < V_p$ | $V_s > V_p$ |
| ☐ D | $V_s = V_p$ | $V_s = V_p$ |

3. Explain why a transformer will not work with a dc supply. [2]
4. A transformer with 200 turns on the primary coil and 20 turns on the secondary coil has a potential difference of 120 V across the primary coil.
   (a) Calculate the potential difference (pd) in the secondary coil.
   (b) Name the type of transformer. [4]
5. A transformer with a pd of 40 V and a current of 0.5 A in the primary coil produces a current of 0.2 A in the secondary coil. Calculate the pd in the secondary coil. [3]

1. *The generator effect.*[1]
2. *B*[1]
3. *A dc supply to the primary coil does not produce a changing magnetic field*[1] *so will not induce a potential difference in the secondary coil*[1].
4. *(a) $V_p / V_s = n_p / n_s$    $120 / V_s = 200 / 20$*[1]  *$V_s = 120 / 10$*[1] *$= 12$ V*[1]
   *(b) It is a step-down transformer*[1]
5. *$V_s \times I_s = V_p \times I_p$    $V_s \times 0.2 = 40 \times 0.5$*[1] *$V_s = 40 \times 0.5 / 0.2$*[1] *$= 100$ V*[1].

# EXAMINATION PRACTICE

01   This question is about magnets and magnetic fields.

    01.1   A bar magnet is hung on a thread so it can move freely.
         Explain what will happen when another bar magnet is brought towards the south pole
         of the hanging magnet as shown.                                                    [2]

    01.2   Name **two** magnetic materials.                                                      [2]

    01.3   Permanent and induced magnets are both made of magnetic materials.
         Describe the differences between a permanent and an induced magnet.                [2]

    01.4   Draw the magnetic field pattern of a bar magnet.                                   [3]

02   This question is about electromagnetism.

    02.1   Draw the magnetic field pattern for a straight wire carrying a current.            [2]

    02.2   Draw the magnetic field pattern for the wire when it is coiled into a solenoid.    [3]

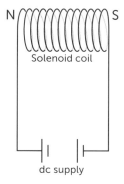

A student investigated factors that affect the strength of an electromagnet by measuring the number of pins the electromagnet picked up each time.
The table shows the number of pins picked up under different conditions.

| Number of coils | Number of cells | | | | | |
|---|---|---|---|---|---|---|
| | 1 | | 2 | | 3 | |
| | No core | Iron core | No core | Iron core | No core | Iron core |
| 5 | 1 | 5 | 4 | 8 | 5 | 10 |
| 10 | 3 | 7 | 2 | 12 | 9 | 17 |
| 15 | 5 | 9 | 7 | 15 | 11 | 20 |
| 20 | 7 | 15 | 9 | 19 | 13 | 25 |

02.3  Which result is anomalous? [1]

02.4  Write three conclusions about the factors that affect the strength of an electromagnet from the results. [3]

02.5  **Physics only:** Explain how the electric bell shown in the diagram works. [5]

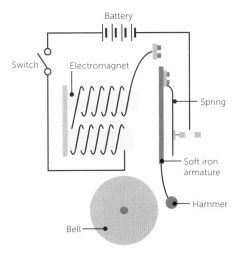

03  **Higher Tier only:** The diagram shows a wire carrying a current at right angles to a magnetic field.

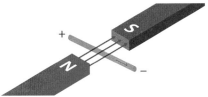

03.1  The wire is 1.8 m long. The current is 1.2 A. The magnetic flux density of the magnetic field is 0.03 T.
Calculate the force exerted on the wire. [2]

03.2  Draw an arrow on the diagram to show the direction of the force. [1]

03.3  A coil of wire carrying a current is placed in a magnetic field.
Explain how the force on the wire causes the coil to rotate. [2]

03.4 **Physics Higher Tier only:** The diagram shows a moving coil headphone.
Explain how a moving-coil headphone converts an electrical signal into a sound wave. [6]

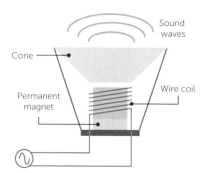

04 **Physics Higher Tier only:** This question is about the generator effect.

04.1 Give **three** ways to increase the size of an induced potential difference in a wire. [3]

04.2 Draw a graph of potential difference against time for two complete rotations of:
(a) an alternator [2]
(b) a dynamo [2]

04.3 Explain how using different commutators produces different currents in an alternator and a dynamo. [4]

04.4 The diagram shows a moving-coil microphone. Explain how it works. [4]

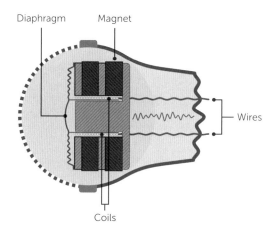

05 **Physics Higher Tier only:** A basic transformer consists of a primary coil and a secondary coil wound on an iron core.

05.1 Give **two** reasons why an iron core is used. [2]

05.2 A transformer steps down a potential difference of 11.04 kV to 230 V.
The transformer has 960 turns in the primary coil. Calculate the number of turns in the secondary coil. [3]

05.3 The transformer has a current of 3.0 A in the primary coil. Calculate the current in the secondary coil. [3]

# OUR SOLAR SYSTEM

## The Sun, the planets and their moons

Stars are not actually living but we describe the time they exist as a life cycle.

The Sun is the star at the centre of our solar system.

Orbiting around the Sun are:

- Eight planets - Mercury, Venus, Earth, Mars, Jupiter, Saturn, Uranus and Neptune.
- Dwarf planets such as Pluto.
- Moons (natural satellites), comets and asteroids.

All of the bodies in the solar system are tiny compared with the Sun. Scale drawings of the solar system are rarely shown because of the large range of sizes and the vast distances involved.

Stars are grouped in **galaxies** which contain many billions of stars. The **universe** contains many billions of galaxies.

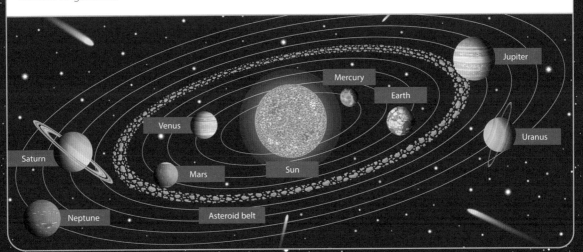

## Formation of a star

Stars including our Sun start life from a nebula which is a cloud of dust and gas. The **nebula** is pulled together by gravitational attraction. The energy transfers cause a massive rise in temperature and nuclear fusion reactions begin. The gas is mainly hydrogen so at this stage hydrogen nuclei fuse to form helium nuclei.

The start of fusion leads to two opposing forces within a star:

- The **inward** force of **gravity** which causes the star to **collapse**.
- The **outward** force from the **energy** of nuclear fusion which causes the star to **expand**.

An equilibrium is reached when these two forces are balanced. The star starts to emit light and is now a stable **main sequence star**. This is the longest stage for a star.

1. Name the galaxy our solar system belongs to. [1]
2. Name the **four** relatively small rocky planets. [1]
3. Name the **four** relatively large gas planets. [1]

   1. *The Milky Way galaxy.*[1]
   2. *Mercury, Venus, Earth and Mars.*[1]
   3. *Jupiter, Saturn, Uranus and Neptune.*[1]

Remember: **Nuclear fusion** is when two light nuclei join together to form one heavier nucleus.

# THE LIFE CYCLE OF A STAR

The **life cycle** of a star describes how a star changes over time from its formation to its end.

### Life pathways

After the **main sequence** stage, one of two paths is taken. This is determined by the **mass** of the star. Smaller stars, like our Sun, exist for billions of years. The more massive stars have a much shorter lifespan of a few million years. Fusion processes during the different stages produce all of the naturally occurring elements in the universe.

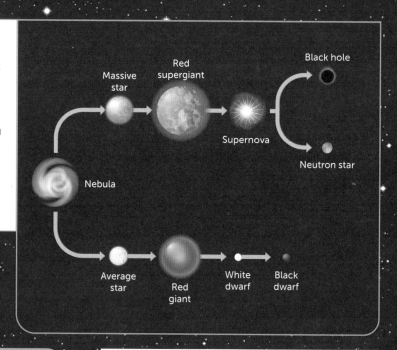

### For a star the size of the Sun

When most of the hydrogen gas has been converted to helium, the inward force of gravity is greater than the outward force from fusion, so the star collapses. This causes a temperature rise and helium nuclei now fuse to form larger elements up to the mass of **iron**. The star then expands and becomes a **red giant**. Fusion stops when all the elements are used up and the star shrinks to be a **white dwarf**. As the fusion has stopped, the star stops releasing energy and cools. The final stage is a **black dwarf**.

Which statements correctly describe what happens during the life cycle of a star?
Tick **two** boxes.      [2]

  ☐ A    All stars end as black holes.

  ☐ B    Elements heavier than iron are produced in white dwarfs.

  ☐ C    Fusion releases energy.

  ☐ D    Higher temperatures are needed for the fusion of heavier elements.

  ☐ E    The temperature decreases when a star collapses.

*C[1], D[1]*

### For a star much more massive than the Sun

For stars much larger than the Sun, a **red super giant** forms when all of the hydrogen is used up. After this, the star explodes as a **supernova**. Elements heavier than iron are then formed because the temperature is high enough for larger nuclei to fuse together. All the elements in the universe are distributed from this massive explosion. Stars up to about four times the mass of the Sun are pulled together by gravity forming a **neutron star** made up of densely packed neutrons. The larger stars go on to become **black holes**.

# ORBITAL MOTION AND SATELLITES

## Orbiting objects

'The Moon' is the name of our moon and 'a moon' is the general term for all moons.

Planets orbit the Sun in elliptical orbits which we consider to be **circular orbits**. A **satellite** is an object that orbits a planet. These are:

- **Natural** satellites, for example, moons.
- **Artificial** (manufactured) satellites sent into space for exploration and communication.

All planets, moons and artificial satellites orbit another object.

**Gravity** provides the force that allows planets, moons and artificial satellites to maintain their circular orbits. Gravity provides a resultant force towards the centre of the orbit, at 90° to the direction of movement. Without gravity, the object would just move in a straight line.

## Speed of orbiting objects        Higher Tier

The velocity of an object is its speed in a given direction.

For an object in orbit, the force of gravity keeps it moving at a constant speed at the same radius from the centre of the orbit. Because the object is always changing direction, the velocity of the orbiting object also changes. This means the orbiting object is also accelerating.

The speed of an orbiting object depends on:

- the force of gravity provided by the planet it orbits around.
- the radius of the orbit.

To **change the speed** of an object in a stable orbit, the **radius** of the orbit must change. A faster-moving object needs a greater force of gravity to keep it in a stable orbit. Planets closer to the Sun move faster than planets further away from the Sun.

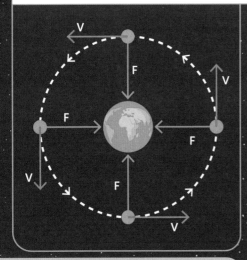

A **stable orbit** means the orbiting object has a constant speed and is a fixed distance (the radius) from the point the object rotates about.

1. Complete the sentences.
   Artificial satellites orbit the _____.        [1]
   Natural satellites orbit the _____.        [1]
   Planets orbit the _____.        [1]

2. Describe the orbit of a geostationary satellite.        [1]

3. **Higher Tier only:** Describe the relationship between the speed and the orbital radius of an object in a stable orbit.        [1]

   1. Earth.[1] Planets.[1] Sun.[1]

   2. The satellite orbits above the same point on Earth at all times.[1]

   3. The faster the speed, the shorter the orbital radius.[1]

# RED-SHIFT

**Red-shift** is the observed increase in the wavelength of light from most distant galaxies. When the spectrum of light from stars is examined, there are black lines where wavelengths have been absorbed. The light waves coming from a distance galaxy appear to be stretched so the observed **wavelength** is **increased**. This moves the black line towards the red part of the spectrum.

### Spectrum from a nearby star

### Spectrum from a galaxy that is far away

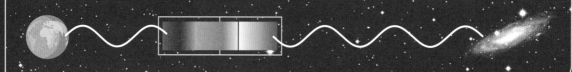

The **further away** the galaxy is, the greater the observed increase in wavelength, which means the galaxy is **receding faster**.

The **Big Bang theory** suggests that the universe began as an extremely small, hot, dense region which started to expand. Red-shift provides evidence for this theory and supports the idea that the universe expanded and continues to expand.

Theory suggests that the rate of expansion should slow down because of gravity. However, since 1998, observations of supernovae suggest that the rate that distant galaxies are receding is actually increasing. This could be explained by the existence of **dark mass** and **dark energy** which is not yet fully understood.

Tick **one** box in each row to show whether each statement is true or false. [3]

| Evidence from red-shift | True | False |
|---|---|---|
| The universe is expanding | | |
| The Big Bang theory is definitely correct | | |
| All galaxies travel at the same speed | | |
| Red-shift is smaller for galaxies that are further away | | |
| The further away a galaxy is the faster it moves | | |
| The faster a galaxy moves the greater the red-shift | | |

*T, F, F, F, T, T.[3] (5 correct.[2] 4 correct.[1])*

# EXAMINATION PRACTICE

**Physics only**

01   The Sun is a star at the centre of our solar system.

01.1   Give **two** other types of body that are part of our solar system. [2]
The life cycle of the Sun has six stages.

**Black dwarf    Main sequence star    Nebula    Protostar    Red giant    White dwarf**

01.2   Put the stages in order from start to finish. [1]

01.3   Give the current stage the Sun is at. [1]

01.4   Explain how an equilibrium in the size of a start is reached during its formation. [3]

01.5   Some stars are much more massive than the Sun. Describe the life cycle of a
massive star after the main sequence stage. [6]

02   The International Space Station (ISS) orbits the Earth in a stable orbit.

02.1   Explain how the ISS stays in its orbit. [2]

02.2   Give **two** uses of artificial satellites. [2]

02.3   **Higher Tier only:** Explain what would happen if the ISS was boosted in to a higher orbit. [2]

03   This question is about red-shift.

03.1   Describe red-shift. [2]

03.2   What is the relationship between red-shift and the movement of a galaxy?
Tick **one** box. [1]

☐ The greater the red-shift, the faster the galaxy moves away from us.

☐ The greater the red-shift, the faster the galaxy moves towards us.

☐ The greater the red-shift, the slower the galaxy moves away from us.

☐ The greater the red-shift, the slower the galaxy moves towards us.

03.3   Give **one** example of a discovery in the universe that cannot be explained yet. [1]

03.4   The diagram shows the spectrum of light from a distant galaxy.
Describe what would happen to the black lines of a galaxy that is much closer to
the Earth. [1]

# EXAMINATION PRACTICE ANSWERS

01  $E_k = \frac{1}{2}mv^2$ [1]

02. 0.05m [1]

03. $E_k = \frac{1}{2}mv^2 = 0.5 \times 0.4 \times 8^2$ [1] = 12.8 [1] J [1] [3]

04.1 $E_p = mgh = 85 \times 9.8 \times 2$ [1] = 1666 J [1] [2]

04.2 As the cyclist falls, energy is transferred from the gravitational potential energy store to the kinetic energy store, [1] and as the cyclist hits the ground the kinetic energy is transferred to the thermal energy store of the air in the tyre, and the surroundings and/or elastic potential energy store of the tyres [1]. [2]

04.3 **Higher Tier only:** $E_p = E_k$ so $E_k = 0.5 \times 85 \times v^2 = 680$ J [1]   $v = \sqrt{[680 /(0.5 \times 85)]}$ [1] = 4.0 m/s² [1] [3]

05  Trend: The number of diesel cars sales went down 2018-19. [1]

Reason: Any one from: diesel is a known air pollutant / some places are banning diesel cars from city centres / Government targets to ban production of new fossil fuel cars by 2030. [1]

Trend:   The number of electric car sales went up 2018-19. [1]

Reason: Any one from: becoming more available / prices are starting to come down / they are becoming cheaper to run / increase in electric car charging points ; no car tax / people are encouraged to buy them as they contribute less to carbon dioxide emissions. [1] [4]

06  Advantage: (any one from) does not emit carbon dioxide / sulfur dioxide / contribute to global warming. [1]

Disadvantage: (any one from) hugely expensive to build/decommission the power station; disposal of radioactive waste; risk of catastrophic accident. [1] [2]

07  Air is a better insulator than brick [1] so having a layer of air reduces the rate that thermal energy can transfer through the walls [1] increasing the thickness of the wall by having two layers also reduces the rate of energy transfer [1]. [3]

08  Efficiency = useful power output / total power input; 0.45 = useful power output / 30;
useful power output = 0.45 × 30 [1] = 13.5W [1] $E = Pt$ = 13.5 × 20 = 270 J [1] [3]

09  The same amount of energy needs to be supplied by both heaters (to warm up the room) [1] and the more powerful heater transfers that energy more quickly [1]. [2]

10.1 The amount of energy in joules needed to raise the temperature of 1 kg of water by 1°C. [1]

10.2 The time the heater is on. [1]

10.3 Thermal energy from the heater was also transferred to heat up the air/beaker/surroundings [1] so reduce this loss by increasing the insulation of the container [1]. [2]

10.4 **Higher Tier only:** 250 g = 0.25 kg [1]; $\Delta E = mc\Delta\theta$; 2100 = 0.25 × 4200 × $\Delta\theta$ [1]; $\Delta\theta$ = 2100 / (0.25 × 4200) [1] = 2 °C [1] [4]

11  *Suggested method: This is an extended response question that should be marked in accordance with the levels based mark scheme on page 170.* [6]
Indicative content:
Put 80cm³ hot water from a kettle into a 100cm³ beaker.[✔]
Take care with hot water[✔]
Place a cardboard lid on the beaker with a hole for a thermometer.[✔]
Insert the thermometer through the lid into the hot water.[✔]
Record the temperature of the water and start the stopwatch.[✔]
Record the temperature of the water every 3 minutes for 15 minutes.[✔]
Repeat for 2 layers of newspaper around the beaker[✔] held in place with an elastic band [✔].
Aim to have the same starting temperature of the hot water.[✔]
Then also repeat for 4, 6 and 8 layers of newspaper.[✔]
Draw cooling curves for each different number of layers.[✔]
The smaller the temperature drop the better the insulation.[✔]

01.1 ammeter, voltmeter. [2]

01.2

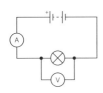

All symbols correct [1] battery, ammeter and lamp in series [1] voltmeter across lamp [1]. [3]

01.3 $V = IR$; $1.5 = 0.5 \times R$ [1] $R = 1.5 / 0.5$ [1] $= 3\ \Omega$ [1] [3]

01.4 250 mA = 0.25 A [1]; $Q = It$; $3.0 \times 10^4 = 0.25 \times t$ [1]; $t = 3.0 \times 10^4 / 0.25$ [1] $= 1.2 \times 10^5$ s [1] [4]

02.1 Variable resistor. [1]

02.2 Changes length of wire connected. [1] Longer wire, higher resistance. [1] [2]

02.3 LDR  [1]

02.4 As the light intensity increases, the resistance decreases. [1] This means the current changes and circuits can be switched on and off. [1] [2]

03.1 $R_r = 2 \times 2.0 = 4.0\ \Omega$ [1] $R_{total} = 2.0 + 4.0 = 6.0\ \Omega$ [2]

03.2 $V = IR$; $12.0 = I \times 6.0$ [1]; $I = 12 / 6 = 2$ A [2]

03.3 Ratio of pd resistor : pd lamp is 2:1. (a) resistor 8 V [1] (b) lamp 4 V [1] (*Could also use V = I R*) [1] / [1]

03.4 Graph C. [1]

04 Direct pd is in one direction only. [1] Alternating pd changes direction.[1] [2]

05 The appliance also becomes live. [1] A person touching it would receive an electric shock because the current goes through them. [1] [2]

06 Step-up transformers increase the pd in the transmission cables [1] and this reduces the current so less energy as dissipated as wasted heat [1]. A step down transformer then reduces the pd back to a safe value for domestic use. [1] [3]

07.1 $P = VI$; $2900 = 230 \times I$ [1] $I = 2900 / 230$ [1] $= 12.6$ A [1] [3]

07.2 The kettle with the higher power rating has a higher current. [1]

07.3 power = (current)$^2$ × resistance [1]

08.1 $E = Pt$; $45\,000 = P \times 70$ [1] $P = 45\,000 / 70$ [1] $= 643$ W [1] [3]

08.2 $E = QV$; $45\,000 = Q \times 230$ [1] $Q = 45\,000 / 230$ [1] $= 196$ C [1] [3]

09.1 The acetate rod has a positive charge and the cloth has a negative charge. [1]

09.2 The charge on the rods is the same [1] so the rods will repel away from each other. [1] [2]

09.3 Radial field [1] with arrows pointing towards electron [1]. (Number of arrows unimportant as long as evenly spread.) [2]

09.4 The person has become charged through friction. [1] As the person touches the door handle, the charge discharges to the door handle [1], which causes a small electric shock. [2]

01.1  0.7364 kg [1]

01.2  $\rho = m / V = 763.4/50.0$ [1] = 15.3 g/cm³ [1] [2]

01.3  $V = 0.02 \times 0.02 \times 0.02$ [1] = $8 \times 10^{-6}$ m³ [1] [2]

01.4  $\rho = m / V$; 5400 mass / $8 \times 10^{-6}$ [1] mass = $5400 \times 8 \times 10^{-6}$ [1] = $4.32 \times 10^{-2}$ kg [1] = 43.2 g [1] [4]

02.1  Energy is transferred from the liquid. [1] The particles lose energy [1] and stop moving freely so just vibrate around fixed positions as a solid. [1] [3]

02.2  The total kinetic and potential energy that the particles in a substance have. [1]

03.1  The amount of energy needed to melt 1kg of a solid to a liquid at a constant temperature. [1]

03.2  $E = mL = 0.25 \times 334\ 000$ [1] = 83 500 J [1] [2]

03.3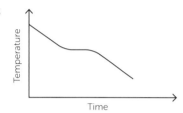

Labelled axes [1], correct shape cooling curve [1]. [2]

04  **Higher only:**
To heat the solid gold to melting point: $\Delta E = mc\Delta\theta = 0.5 \times 130 \times 64$ [1] = 4160 J [1]
To melt the gold: $E = 0.5 \times 63\ 000$ [1] = 31 500 J [1]
Total energy is 4160 + 31 500 = 35660 J [1] [5]

05  Increases. [1]

06.1  **Physics only:** $p_1V_1 = p_2V_2$; $p \times 0.11 = 25 \times 0.05$ [1] $p = 1.25/ 0.11$ [1] = 11.4 Pa [1] [3]

06.2  **Physics (Higher) only:** Work is done to compress the gas in the piston [1] so energy is transferred and the internal energy of the gas particles increases [1] so the temperature of the gas in the piston increases [1]. [3]

01.1  0.04 nm [1]

01.2  $2 \times 4.0 \times 10^{-11}$ m = $8.0 \times 10^{-11}$ m [1]

02.1  9 protons [1] 10 neutrons [1] 9 electrons[1] [3]

02.2  Atomic number 13 [1]   Mass number 27 [1] [2]

02.3  They have different numbers of neutrons. [1]

03.1  The atom is a ball of positive charge [1] with negative electrons embedded in it. [1] [2]

03.2  All three lines correct [2], 1 correct line [1] [2]

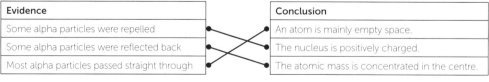

| Evidence | Conclusion |
|---|---|
| Some alpha particles were repelled | An atom is mainly empty space. |
| Some alpha particles were reflected back | The nucleus is positively charged. |
| Most alpha particles passed straight through | The atomic mass is concentrated in the centre. |

04.1  240 / 2 = 120; 120 / 2 = 60 so 2 half-lives is 14 years [1] so one half life is 7 years [1]. [2]

04.2 1 hour is 2 half-lives. $\frac{1}{2} \times \frac{1}{2} = \frac{1}{4}$ kg [1] ratio = 4.0 : 16.0 = 1:4 [1]. [2]

04.3 The longer the half-life is, the longer the radioactive decay will continue at a higher level [1] so will therefore be more of a hazard for longer. [1] [2]

05.1 Range in air – gamma, beta, alpha [1]. Ionising power – alpha, beta, gamma [1]. [2]

05.2 The source for the smoke detector needs to be ionising [✓] and safe [✓]. Nickel-63 is a beta source which is less ionising than alpha [✓] and also passes through skin so is unsafe. [✓] The gamma source cobalt-60 is only weakly ionising [✓] and penetrates a long way so is unsafe. [✓] Alpha radiation is strongly ionising [✓]. It does not penetrate outside the smoke alarm [✓]. So, the best choice is one of the two alpha sources [✓] The one with the longest half-life [✓] will be more appropriate as it will last longer [✓]. Half of the thorium-228 is decayed after just under 2 years [✓] and will need to be changed too often [✓]. So the most appropriate source for the smoke alarm is Americium-241 [✓] as it is an ionising source, has a long half life and can be used safely in the home as long as it is not ingested. [✓]

*This extended response question should be marked holistically in accordance with the levels-based mark scheme on page 170.* [6]

05.3 $^{241}_{95}\text{Am} \longrightarrow ^{237}_{93}\text{Np} + ^{4}_{2}\text{He}$ [1]  ⌐[1]  ⌐[1] [3]

05.4 $^{63}_{28}\text{Ni} \longrightarrow ^{63}_{29}\text{Cu} + ^{0}_{-1}\text{e}$ [1]  ⌐[1]  ⌐[1] [3]

**Physics only**

06.1 Background radiation is the radiation from both natural and man-made sources that is around us all the time. [1]

06.2 Any one from each list:
Natural sources – Rocks / buildings /cosmic rays / food and drink. [1]
Man-made sources – Nuclear weapons / nuclear accidents / medical treatments / nuclear waste. [1] [2]

06.3 The half-life needs to be long enough so that enough measurements can be taken in the body [1] and short enough so the hazard level reduces quickly when it is no longer needed. [1]

06.4 Nuclear radiation can ionise cells and damage them so they mutate into cancer cells [1] and high energy gamma rays can be targeted on the cells to kills them [1]. [2]

07.1

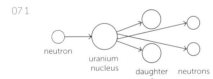

neutron    uranium nucleus    daughter nucleus    neutrons

Neutron hitting uranium. [1]
Uranium splitting into 2 daughter nuclei. [1]
Further neutrons emitted. [1] [3]

07.2 Nuclear fusion is the joining together of two light nuclei to make one heavier nucleus. [1]

## Topic 5

01.1 Displacement. [1]

01.2 $F_y$ is in the opposite direction to $F_x$ [1] and has twice the magnitude $F_x$. [1] [2]

01.3 There is resultant force acting to the right [1] so the box will accelerate to the right. [1] [2]

01.4 Momentum is mass multiplied by velocity. [1] Velocity a vector quantity so momentum will have both magnitude and direction/also be a vector quantity. [1] [2]

01.5 Any **two** from: magnetic, electrostatic, gravitational. [2]

01.6 Weight is the force acting on a mass due to gravity. [1]

01.7 $W = m \times g = 1.6 \times 9.8$ [1] $= 15.7$ N [1] [2]

01.8   Newtonmeter. [1]

02.1   Resultant force = 2 + 3 – 5 [1] = 0 N [1] [2]

02.2   Forces labelled correctly. [1] Right arrow longer than left arrow. [1] Weight and reaction arrows equal. [1] [3]

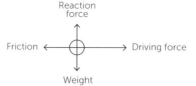

02.3   Horizontal force 2.2 N [1] vertical force 2 N [1] [2]

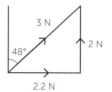

02.4   Scale drawing of original forces [1] resultant force 2150 N [1] at 22° to the horizontal [1]. [3]

03.1   Work done = force × distance. [1]

03.2   $(W = F_s)\ E_p = mgh$ [1]
Work done is against gravity so $F = mg$ [1]
The distance moved along the line of action of the force, so $s = h$ [1]
so work done $(= F_s) = m \times g \times h = E_p$ gained by the mass. [1] [4]

03.3   $W = F_s$; 6600 = 1200 × s [1]; s = 6600/1200 [1] = 5.5 m [1] [3]

03.4   500 Nm = 500 J [1] = 0.5kJ [1] [2]

04.1   Force = spring constant × extension. [1]

04.2   $F = ke$; 2.1 = k × 0.07 [1]; k = 2.1/0.07 [1] = 30 [1] N/m [1] [4]

04.3   3.5 cm = 0.035 m [1]; $E_e = 0.5ke^2$ = 0.5 × 500 × 0.035 × 0.035 [1] = 0.31 J [1] [3]

04.4   e.g. 2.4 – 0.2 = 2.2 cm [1]

04.5   5 Correctly plotted points.[2] Line of best fit.[1] [3]

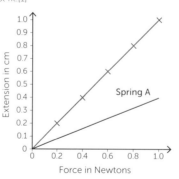

04.6   (k = 1 / gradient). gradient = 0.4/10 [1] k = 25 N/cm [1] [2]

05.1   $M = Fd$ = 9.0 × 0.3 [1] = 2.7 Nm [1] [2]

05.2   The total anticlockwise moment equals the total clockwise moment. [1]

05.3   Total anticlockwise moment = (0.5 × 0.2) + (0.4 × 0.1) = 0.14 Nm [1]
Total clockwise moment = 0.14 = weight of star × 0.2 [1]
weight of star = 0.14/0.2 [1] = 0.7 N [1] [4]

06.1   $P = F/A$ = 25/ 0.2 [1] = 125 [1] Pa [1] [3]

06.2   Atmospheric pressure is caused by air molecules [1] colliding with a surface. [1] [2]

06.3 29 400 = h × 1000 × 9.8 [1] h = 29400/(1000 × 9.8) [1] = 3.0 m [1]  [3]

06.4 The diver experiences a greater pressure on the front than on the back [1] because pressure increases with depth in the water. [1] This means there is a resultant upwards force, or upthrust, on the diver. [1]  [3]

07.1 Displacement is 5.5 km [1] at 45° to the horizontal (or north east) [1]  [2]

07.2 3.6 km = 3600 m [1]; s = vt; 3600 = v × 150 [1]; v = 3600/150 [1] = 24 m/s [1].  [4]

07.3 Graph drawn with distance on vertical axis and time on the horizontal axis, with sensible scale [1].
Points plotted correctly. [1] Point joined with a straight line [1].  [3]

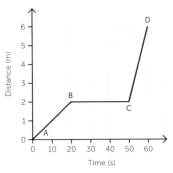

07.4 s = vt; 6 = v × 60 [1]; v = 6/60 [1] = 0.1 m/s [1]  [3]

07.5  [1]

08.1 a = Δv/t; 1.5 = Δv /8 [1]; Δv = 1.5 × 8 [1] = 12 m/s [1]  [3]

08.2 $v^2 - u^2 = 2as$; $9^2 - 5^2 = 2 \times 0.5 \times s$ [1]; s = (81 – 25) / 1.0 [1] = 56 m [1]  [3]

08.3 A is travelling at a constant velocity with zero acceleration. [1]
The velocity of B is increasing, with constant acceleration. [1]
The velocity of C is also increasing, but with a greater constant acceleration than B. [1]  [3]

08.4 acceleration = gradient = 4/20 [1] = 0.2 m/s² [1]  [2]

08.5 distance = area under graph = (0.5 × 20 × 4) + ((35 – 20) × 4) = 40 + 60 [1] = 100 m [1]  [2]

08.6 Velocity and time labelled on axes [1] graph line initially rises then starts to level off [1] horizontal line labelled terminal velocity [1].  [3]

09.1 Inertia.  [1]

09.2 A resultant force must act on the object.  [1]

09.3 The object will continue to move to the right at a constant speed.[1] There will be a resultant force upwards so the object will also accelerate upwards.[1]  [2]

09.4 F = ma; 250 = m × 12.5 [1]; m = 250/12.5 [1] = 20 kg [1].  [3]

09.5 There is a resultant force (of 20 N) to the right [1], so the box will accelerate to the right [1].  [2]

09.6 The force on the trolley is provided by the hanging masses. [1] As the effect of mass is being investigated, the force must be kept constant. [1]  [2]

09.7 Any **one** from: protect people/floor from falling masses; take care with any electrical equipment used.  [1]

09.8 Two objects exert equal and opposite forces on each other when they interact.  [1]

10.1 Stopping distance = thinking distance + braking distance.  [1]

10.2 At 40 mph, the speed and thinking distance double compared with 20 mph.[1] The braking distance is 4 times bigger. [1]  [2]

10.3 Reaction time increases.  [1]

10.4 Any **two** from: wet/icy road surface; worn tyres ; worn brakes. [2]

10.5 A large deceleration means a greater braking force [1] so more work is done transferring the kinetic energy of the vehicle to thermal energy of the brakes [1] so the temperature of the brakes rises rapidly with no time to cool [1]. [3]

11.1 $p = mv = 3.5 \times 0.4$ [1] $= 1.4$ kg m/s.[1] [2]

11.2 For a closed system, the total momentum before an event equals the total momentum after an event. [1]

11.3 $m_1v_1 + m_2v_2 = (m_1 + m_2)v_3$; $(15\,000 \times 8) + (m \times 3) = (15\,000 + m) \times 6$ [1]; $120\,000 + 3m = 90\,000 + 6m$; $m = (120\,000 - 90\,000)\,/3$ [1] $= 10\,000$ kg [1] [3]

11.4 $F = m\Delta v/\Delta t = (2.5 \times 15)\,/\,0.2$ [1] $= 180$ N [1] [2]

## Topic 6

01.1 [3]

Compression [1]   Rarefaction [1]

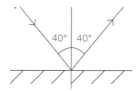

Wavelength [1]

01.2 $T = 1/f$; $0.4 = 1/f$ [1]; $f = 1/0.4$ [1] $= 2.5$ [1] Hz[1] [4]

01.3 $v = f\lambda$; $160 = 8 \times \lambda$ [1]; $\lambda = 160/8$ [1] $= 20$ m [1] [3]

01.4 *Suggested method: This is an extended response question that should be marked in accordance with the levels based mark scheme on page 170.*
The students stand together in front of a large wall.[✔] Measure the distance from the wall [✔] using a long tape measure [✔] and record the distance in metres. [✔] Make the distance as large as possible [✔] to increase the time taken for the sound to travel. [✔] One student makes a sound by banging two blocks together. [✔] The other student starts timing with a stopwatch as the sound is made [✔] stops timing when the echo is heard and records the time in seconds. [✔] Repeat this several times.[✔] Remove any anomalies [✔] and calculate a mean time.[✔] Swap the students around and repeat the experiment [✔] to reduce the effect of their different reaction times. [✔] The distance the sound has travelled is two times the distance to the wall. [✔] Calculate the speed of sound using speed = distance/time. [✔] [6]

02.1 wavelength = length of tank/number of waves = 0.56/28 [1] = 0.02 m. [1] [2]

02.2 $(2.9 + 3.3 + 3.2 + X)/4 = 3.2$[1]; $X = (3.2 \times 4) - (2.9 + 3.3 + 3.2) = 3.4$ (Hz). [1] [2]

02.3 $v = f\lambda = 3.2 \times 0.075$ [1] = 0.24 m/s. [1] [2]

02.4 A video can be used to determine both frequency and wavelength [1] as long as the timer is also recorded [1]. A photograph can help determine wavelength but not frequency. [1] [3]

02.5 $\lambda = 2 \times 0.75/5$ [1] = 0.3 m [1]
$v = f\lambda = 12 \times 0.3$ [1] = 3.6 m/s [1] [4]

03.1 The wave can be absorbed, reflected and transmitted. [1 mark for two, 2 marks for all three] [2]

03.2 normal drawn [1]; $i = r = 40°$ [1] [2]

40°   40°

03.3 Correct angles for material 1. [1] Correct angles for material 2. [1] [2]

|  | Material 1 | Material 2 |
|---|---|---|
| Angle of incidence | 50° | 50° |
| Angle of reflection | 50° | 50° |
| Angle of refraction | 25° | 15° |

03.4 Whatever the material, angle of incidence = angle of reflection. [1]
For the same angle of incidence, the angle of refraction varies for different materials. [1]  [2]

04.1 Any **two** from: Transverse waves/ transfer energy from a source to an absorber / can travel through a vacuum /
travel at the same speed through air or a vacuum.  [2]

04.2 Ray labelled refracted ray with arrow as shown [1] angle of refraction labelled which is less than the angle of incidence [1].  [2]

```
                              Air   Water
                                    Refracted ray

                        Angle of
                        incidence
         Normal

                                    Angle of
                                    refraction

              Incident ray
```

04.3 The waves slow down as they move from deep to shallow water (shorter wavelength) [1]. As the wavefront touches the
boundary the speed of that part slows down whilst the rest of the wavefront continues at the faster speed until it also
meets the boundary and slows down. [1] This causes the wavefront to change direction and travel through the shallow
water at a different angle. [1]  [3]

04.4 Independent variable: the type of surface. [1] Dependent variable: the amount of infrared radiation emitted. [1]  [2]

04.5 Any one from: the distance of the IR detector from each surface / the temperature of the surface / the time taken to
measure the amount of radiation emitted.  [1]

04.6 Lines drawn linking: A to shiny black; B to shiny silver; C to matt white; D to matt black [2] Allow [1] for 2 or 3 correct.  [2]

04.7 Gamma rays.  [1]

04.8 An electric field is created around a wire when a current flows through it.[1] When the current changes, the electric field
changes.[1] The changing current produces radio waves in the electric field.[1].  [3]

04.9 Any two from: Electrical heaters / cooking food / infrared cameras.  [2]

04.10 Radio waves are not easily absorbed in the atmosphere so can travel long distances. [1] Their energy is not high
enough to cause any harm or other effects.[1]  [2]

05.1 20 Hz to 20 000 Hz.  [1]

05.2 Sound waves travel by particles passing on vibrations. [1] In a solid the particles are closer together so can transfer the
vibrations more quickly. [1]  [2]

05.3 P-waves are longitudinal and S-waves transverse [1] P-waves can pass through solids and liquids, S-waves can pass
through solids, but not liquids. [1]  [2]

05.4 distance to the bottom and back = $vt$ = 1500 × 2.5 [1] = 3750 m [1] so depth of water = 3750/2 = 1875 m[1].  [3]

06.1 Convex lens - real or a virtual image, concave - virtual image only [1] convex lens - light converges,
concave lens - light diverges. [1]  [2]

06.2 2.6 cm = 26 mm [1]; magnification = image height / object height = 26/0.052 [1] = 500 [1].  [3]

06.3 Correct symbol for lens [1] at least two light rays diverging [1].  [2]

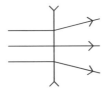

06.4 Blue [1] black [1] blue [1].  [3]

07.1 A perfect black body is an object that absorbs all of the radiation incident on it [1] and reflects and transmits no radiation [1].  [2]

07.2 As the temperature of an object increases it emits more radiation with shorter wavelengths [1] and with a greater intensity [1].  [2]

07.3 It increases.  [1]

01.1 The hanging magnet will move away when the south pole is brought towards the south pole [1] and move towards it when the north pole is brought towards the south [1]. [2]

01.2 Any **two** from: cobalt, iron, nickel, steel. [2]

01.3 A permanent magnet produces its own magnetic field whereas an induced magnet becomes a magnet when it is placed in a magnetic field.[1] A permanent magnet can attract and repel whereas an induced magnet can only attract. [1] [2]

01.4 At least 2 loops above and below equally spaced [1] field lines closer together at both poles [1] arrows from N to S [1]. [3]

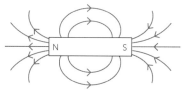

02.1 Concentric circles [1] clockwise direction [1]. [2]

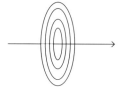

02.2 At least 2 loops above and below equally spaced [1] field lines closer together at both poles [1] field direction labelled [1]. [3]

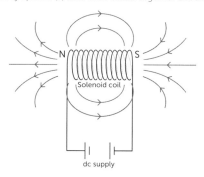

02.3 2 (pins) (10 coils, 2 cells, no core) [1]

02.4 The strength of the electromagnet increases when: there are more coils [1] there is an iron core [1] there are more cells/is a greater current/voltage [1]. [3]

02.5 Closing the switch turns on the electromagnet. [1] The magnet attracts the soft iron armature, and the hammer hits the bell. [1] This breaks the circuit and the electromagnet switches off. [1] The armature is released, and the circuit reconnects. [1] This cycle repeats until the electromagnet is switched off with the switch. [1] [5]

03.1 $F = B I l = 0.03 \times 1.2 \times 1.8$ [1] = 0.065 N [1]. [2]

03.2 Upwards arrow from conductor labelled F. [1]

03.3 The current travels in opposite directions on each side of the coil of wire [1] so the force acting in each side will be in opposite directions producing a moment on the coil. [1] [2]

03.4 *This is an extended response question that should be marked in accordance with the levels based mark scheme on page 170.*

The current in the coil produces a magnetic field in the coil [✓]. This field interacts with the magnetic field produced by the permanent magnet. [✓] They exert a force on each other [✓] and this force moves the cone. [✓] As the current changes [✓] so the strength of the force on the cone changes [✓] and this causes the cone to move in and out. [✓] This vibration causes air molecules to vibrate [✓] and the movement of the air molecules produces the pressure variations that are the sound waves [✓]. [6]

04.1 Increase the strength of the magnetic field the wire is moving relative to [1] increase the speed the wire moves through the field [1] shape the wire into a coil [1].  [3]

04.2 (a) alternator : alternate above and below zero line [1] two whole waves as shown [1].  [2]
     (b) dynamo: all above zero line [1] 2 whole waves as shown [1].  [2]

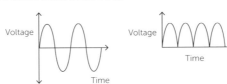

04.3 An alternator uses two complete metal rings called slip ring commutators. [1] A dynamo uses one split metal ring with two gaps in it, called a split ring commutator [1]. In an alternator, the slip ring means that the direction of the current reverses when the coil goes through a complete cycle.[1] In a dynamo, the connection to the coil is reversed every half-cycle, so the current is always in the same direction as the coil rotates. [1]  [4]

04.4 Vibrations in the air from sound waves cause the diaphragm to vibrate.[1] The coil is attached to the diaphragm and so this moves backwards and forwards in the magnetic field of the magnet. [1] The movement of the coil in the magnetic field means that the generator effect induces a potential difference in the coil [1]. If the coil is connected to a complete circuit, a current flows. [1]  [4]

05.1 Iron is easily magnetised. [1] increases the strength of the magnetic field. [1]  [2]

05.2 $V_p / V_s = n_p / n_s$; 11 040/230 = 960 / $n_s$ [1] $n_s$ = (960 × 230) / 11 040 [1] = 20 turns. [1]  [3]

05.3 $V_s × I_s = V_p × I_p$ ; 230 × $I_s$ = 11 040 × 3 [1]; $I_s$ = (11 040 × 3) / 230 [1] = 144 A [1]  [3]

## Topic 8

01.1 Any **two** from: Planets/ dwarf planets/ moons/ asteroids/comets.  [2]

01.2 Nebula, protostar, main sequence star, red giant, white dwarf, black dwarf.  [1]

01.3 Main sequence stage.  [1]

01.4 An inward gravitational force pulls the dust and gas in a nebula together [1] eventually, the temperature and pressure is high enough so that nuclear fusion reactions occur and there is an outwards expansion force due to the fusion energy [1] An equilibrium is reached when these two opposing forces are balanced. [1]  [3]

01.5 *This is an extended response question that should be marked in accordance with the levels based mark scheme on page 170.*

When all of the hydrogen has fused [✓] and converted to helium [✓], the inward force of gravity is greater than the outward force from fusion [✓] so the star collapses [✓]. This causes a temperature rise [✓] and helium nuclei now fuse to form larger elements up to the mass of iron. [✓] The star then expands and forms a red super giant. [✓] After this, the star explodes as a supernova. [✓] This is where elements heavier than iron are formed [✓] because the temperature is high enough for larger nuclei to fuse together [✓]. Stars up to about four times the mass of the Sun are pulled together by gravity [✓] forming a neutron star made up of densely packed neutrons [✓]. The larger stars go on to become black holes. [✓]  [6]

02.1 The force of gravity from the Earth acts on the ISS. [1] The resultant force towards the centre of the Earth keeps it in its orbit. [1]  [2]

02.2 Any **two** from: GPS, space exploration, weather monitoring, observation.  [2]

02.3 The radius of its orbit will increase [1], so its speed would decrease (for it to maintain a stable orbit) [1]  [2]

03.1 Red-shift is when light from a galaxy moving away is shifted to the red end of the spectrum [1] so that the observed wavelength of the light coming from it increases/frequency decreases [1].  [2]

03.2 The greater the red shift, the faster the galaxy moves away from us.  [1]

03.3 Dark mass / dark energy.  [1]

03.4 The black lines in the spectrum would move to the right/move away from the red end of the spectrum.  [1]

# LEVELS BASED MARK SCHEME FOR EXTENDED RESPONSE QUESTIONS

## What are extended response questions?

**Extended response questions** are worth 4, 5 or 6 marks. These questions are likely to have command words such as 'compare', describe', 'explain' or 'evaluate'. You need to write in continuous **prose** when you answer one of these questions. This means you must write in full sentences, not bullet points, and organise your answer into paragraphs.

You may need to bring together skills, knowledge and understanding from two or more areas of the specification. To gain full marks, your answer needs to be logically organised, and ideas linked to give a sustained line of reasoning.

Some extended response questions may involve calculations. These need two or more steps that must be done in the right order. These questions will include the command words 'calculate' or 'determine'.

## Marking

Written answers are marked using 'levels of response' mark schemes. Examiners look for relevant points (indicative content) and also use a best fit approach. This is based on your answer's overall quality and its fit to descriptors for each level. Extended response calculations give marks for each step shown.

## Example level descriptors

Level descriptors vary, depending on the question being asked. Level 3 is the highest level and Level 1 is the lowest level. No marks are awarded for an answer with no relevant content. The table gives examples of the typical features that examiners look for.

| Level | Marks | Descriptors for a method | Descriptors for an evaluation |
|-------|-------|--------------------------|-------------------------------|
| 3 | 5–6 | The method would lead to a valid outcome. All the key steps are given, and they are ordered in a logical way. | The answer is detailed and clear. It includes a range of relevant points that are linked logically. The answer uses relevant data that may be given in the question. A conclusion is made that matches the reasoning in the answer. |
| 2 | 3–4 | The method might not lead to a valid outcome. Most of the key steps are given, but the order is not completely logical. | The answer is mostly detailed but not always clear. It includes some relevant points with an attempt at linking them logically. Data may not be used fully. A conclusion is given that may not fully match the reasoning given. |
| 1 | 1–2 | The method would not lead to a valid outcome. Some key steps are given, but they are not linked in a clear way. | The answer gives separate, relevant points. Uses little or no data that may be given in the question. The points made may be unclear. If a conclusion is given, it may not match the reasoning given in the answer. |

# COMMAND WORDS

A **command word** in a question tells you what you are expected to do.

## The structure of a question

You should see one command word per sentence, with the command word coming at the start. A command word might not be used, however, if a question is easier to follow without one. In these cases, you are likely to see:

- What ...?
- Why ...?
- How ...?

| Command word | What you need to do |
|---|---|
| Balance | Add correct balancing numbers to a nuclear equation. |
| Calculate | Use the numbers given to work out an answer. |
| Choose | Select from a range of options. |
| Compare | Write about **all** the similarities and/or differences between things. |
| Complete | Complete sentences by adding your answers in the spaces provided. |
| Define | Give the meaning of something. |
| Describe | Recall a fact, event or process accurately. |
| Design | Describe how something will be done, such as a practical method. |
| Determine | Use the data or information given to you to obtain an answer. |
| Draw | Produce a diagram, or complete an existing diagram. |
| Estimate | Work out an approximate value. |
| Evaluate | Use your knowledge and understanding, and the information supplied, to consider evidence for and against something. You must include a reasoned judgement in your answer. |
| Explain | Give the reasons why something happens, or make something clear. |
| Give, name, write | Only write a short answer, commonly just a single word, phrase or sentence. |
| Identify | Name or point out something. |
| Justify | Support your answer using evidence from the information given to you. |
| Label | Add the correct words or names to a diagram. |
| Measure | Use a ruler or protractor to obtain information from a photo or diagram. |
| Plan | Write a method. |
| Plot | Mark data points on a graph. |
| Predict | Write a likely outcome of something. |
| Show | Give structured evidence to come to a conclusion. |
| Sketch | Make an approximate drawing, such as a graph without axis units. |
| Suggest | Apply your knowledge and understanding to a new situation. |
| Use | You **must** base your answer on information given to you, otherwise you will not get any marks for the question. You might also need to use your own knowledge and understanding. |

# PHYSICS EQUATIONS

**Mathematical skills account for at least 30% of the marks in the Physics exams and 20% overall in Trilogy.**

Calculation questions in physics usually involve an equation. Most equations need to be learnt and then recalled and applied in the exam. Others need to be selected from a sheet given in the exam.

Some equations are **higher tier only** and some are **physics only**.

All the equations you are expected to learn, select and apply are given here.

**Physics only** (P)

**Higher only** (H)

### Select and apply

| Word equation | Symbol equation | | |
|---|---|---|---|
| pressure due to column of liquid = height of column × density of liquid × gravitational field strength (g) | $p = h \rho g$ | (P) | (H) |
| (final velocity)² − (initial velocity)² = 2 × acceleration × distance | $v^2 - u^2 = 2\,a\,s$ | | |
| force = $\dfrac{\text{change in momentum}}{\text{time taken}}$ | $F = \dfrac{m\,\Delta v}{\Delta t}$ | (P) | (H) |
| elastic potential energy = 0.5 × spring constant × (extension)² | $E_e = \frac{1}{2}\,k\,e^2$ | | |
| change in thermal energy = mass × specific heat capacity × temperature change | $\Delta E = m\,c\,\Delta\,\theta$ | | |
| period = $\dfrac{1}{\text{frequency}}$ | $T = \dfrac{1}{f}$ | | |
| magnification = $\dfrac{\text{image height}}{\text{object height}}$ | | (P) | |
| force on a conductor (at right angles to a magnetic field) carrying a current = magnetic flux density | $F = B\,I\,l$ | (H) | |
| thermal energy for a change of state = mass × specific latent heat | $E = m\,L$ | | |
| $\dfrac{\text{potential difference across primary coil}}{\text{potential difference across secondary coil}} = \dfrac{\text{number of turns in primary coil}}{\text{number of turns in secondary coil}}$ | $\dfrac{V_p}{V_s} = \dfrac{n_p}{n_s}$ | (P) | (H) |
| potential difference across primary coil × current in primary coil = potential difference across secondary coil × current in secondary coil | $V_p\,I_p = V_s\,I_s$ | (H) | |
| For gases: pressure × volume = constant | $p\,V = constant$ | (P) | |

## Recall and apply

| Word equation | Symbol equation |
|---|---|
| weight = mass × gravitational field strength ($g$) | $W = m\,g$ |
| work done = force × distance (along the line of action of the force) | $W = F\,s$ |
| force applied to a spring = spring constant × extension | $F = k\,e$ |
| moment of a force = force × distance (normal to the direction of force) | $M = F\,d$  **P** |
| pressure = $\dfrac{\text{force normal to a surface}}{\text{area of that surface}}$ | $p = \dfrac{F}{A}$  **P** |
| distance travelled = speed × time | $s = v\,t$ |
| acceleration = $\dfrac{\text{change in velocity}}{\text{time taken}}$ | $a = \dfrac{\Delta v}{t}$ |
| resultant force = mass × acceleration | $F = m\,a$ |
| momentum = mass × velocity | $p = m\,v$  **H** |
| kinetic energy = 0.5 × mass × (speed)$^2$ | $E_k = \dfrac{1}{2}m\,v^2$ |
| gravitational potential energy = mass × gravitational field strength ($g$) × height | $E_p = m\,g\,h$ |
| power = $\dfrac{\text{energy transferred}}{\text{time}}$ | $P = \dfrac{E}{t}$ |
| power = $\dfrac{\text{work done}}{\text{time}}$ | $P = \dfrac{W}{t}$ |
| efficiency = $\dfrac{\text{useful output energy transfer}}{\text{total input energy transfer}}$ | |
| efficiency = $\dfrac{\text{useful power output}}{\text{total power input}}$ | |
| wave speed = frequency × wavelength | $v = f\,\lambda$ |
| charge flow = current × time | $Q = I\,t$ |
| potential difference = current × resistance | $V = I\,R$ |
| power = potential difference × current | $P = V\,I$ |
| power = (current)$^2$ × resistance | $P = I^2\,R$ |
| energy transferred = power × time | $E = P\,t$ |
| energy transferred = charge flow × potential difference | $E = Q\,V$ |
| density = $\dfrac{\text{mass}}{\text{volume}}$ | $\rho = \dfrac{m}{V}$ |

# MATHS SKILLS FOR SCIENCE

## Standard form

Standard form is a way of writing very large or very small numbers and is written as:

$$A \times 10^B$$

- A is a number greater than or equal to 1 and less than 10
- B is any integer (negative or positive whole number)

**Examples**

| Number | Standard form |
|---|---|
| 0.0050 61 | $5.61 \times 10^{-3}$ |
| 170 000 000 | $1.7 \times 10^8$ |
| 0.012 03 | $1.203 \times 10^{-2}$ |
| 8 040 000 | $8.04 \times 10^6$ |

## Rounding to *n* decimal places

When rounding to *n* decimal places (dp):

- look at the nth decimal place
- if the next digit is 5 or more, round up by increasing the preceding digit by one
- if it is 4 or less, keep the preceding digit the same

For example, 0.365 rounded to 2 dp is 0.364 is 0.36.

0.8675 rounded to 1 dp is 0.9

## Rounding to *n* significant figures

- Look at the first non-zero digit, go *n* – 1 digits to the right, and follow the rules for rounding to *n* decimal places
- Fill any places after it with a zero and stop when you reach the decimal point.

**Examples**

| | 1 sf | 2 sf | 3 sf |
|---|---|---|---|
| 9375 | 9000 | 9400 | 9380 |
| 56.27 | 60 | 56 | 56.3 |
| 0.003684 | 0.004 | 0.0037 | 0.00368 |

## Units and conversions

Quantities in one unit can be converted into a different unit and the size of the measurement will still be the same.

Units need to be correct when using equations, it is usually the SI unit.

A common conversion is to change time from hours and minutes to the SI unit of seconds:

seconds = hours × 60 × 60

seconds = minutes × 60

Other conversions depend on the prefix of the unit

**Common prefixes**

| | $10^B$ | Prefix | Symbol |
|---|---|---|---|
| 1 000 000 000 | $10^9$ | Giga- | G |
| 1 000 000 | $10^6$ | Mega- | M |
| 1 000 | $10^3$ | Kilo- | k |
| 0.01 | $10^{-2}$ | Centi- | c |
| 0.001 | $10^{-3}$ | Milli- | m |
| 0.000 001 | $10^{-6}$ | Micro- | μ |
| 0.000 000 001 | $10^{-9}$ | Nano- | n |

**Example conversions for the SI unit of metres (m)**

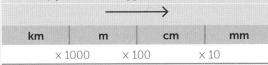

| Multiply to convert bigger units to smaller units | | | |
|---|---|---|---|
| km | m | cm | mm |
| | × 1000 | × 100 | × 10 |

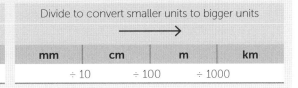

| Divide to convert smaller units to bigger units | | | |
|---|---|---|---|
| mm | cm | m | km |
| | ÷ 10 | ÷ 100 | ÷ 1000 |

## Using data

The **mean** is an average of a set of values.

To calculate the mean, add up all the values and divide this total by the number of values there are.

$$\text{mean} = \frac{\textbf{total sum of the values}}{\textbf{how many values there}}$$

## Graphs

The most commonly used graph in physics is a **line graph** although you could see **bar** charts and **pie** charts too.

A line graph shows the relationship between two **continuous** variables.
- the **independent** variable is plotted on the horizontal **x-axis**.
- the **dependent** variable is plotted on the vertical **y-axis**.

| Plotting a line graph | Drawing a line of best fit |
|---|---|
| • Look at the **range** of values you need to plot on each axis.<br><br>• Choose appropriate **scales** for the small and large squares.<br><br>• Intervals such as 1, 2, 5, and multiples such as 10 or 100 are good to use. Do not use other intervals such as 3, 6 or 9.<br><br>• Make sure each axis uses at least half of the height or width of any given grid.<br><br>• Label both axes with the correct variable and unit.<br><br>• To plot (x, y) find the value on the x-axis, then go up to the value on the y-axis.<br><br>• Use a sharp pencil to plot each point as a small x, accurate to ± 1 small square. | A line of best fit is an indication of the relationship between two variables from experimental data. Lines of best fit can be **straight** or **curved**.<br><br>• Ignore any clearly **anomalous** points.<br><br>• Use a sharp pencil.<br><br>• Draw the line of best fit through most of the points with equal numbers of points above and below the line.<br><br>• Use a transparent ruler for **straight** lines so you can see all the points.<br><br>• Draw a **curved** line free hand as a smooth curve, not dot to dot with a ruler.<br><br>• Avoid drawing double lines. |

# Information from graphs

## Directly proportional

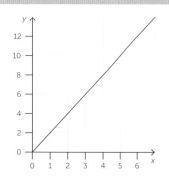

- Straight line graph
- Through the origin
- When x doubles, y doubles
- So x and y are directly proportional to each other

## Inversely proportional

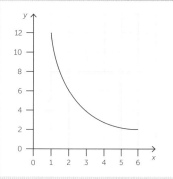

- Curved graph
- When x doubles, y halves
- So x and y are inversely proportional to each other.

## Determining a gradient from a straight line graph

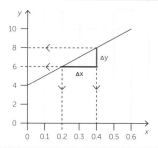

- Choose 2 points on the line which have easy to read values.
- Draw a right-angled triangle from these 2 points
- Determine the values of Δ**y** and Δ**x**
- Gradient = Δ**y** / Δ**x**
- y = mx + c where m is the gradient and c is the point the line intercepts the y axis

### Example

Δy = 8 − 6 = 2                    Δx = 0.4 − 0.2 = 0.2

gradient = 2 / 0.2 = 10     y = 10x + 4

## Determining a gradient from a curved graph

**Higher Tier only**

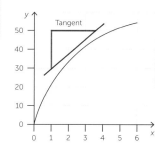

- A **tangent** is a straight line that touches a curve at one point
- Draw a tangent at the point (e.g. x = 2) on the line where you want to find the gradient
- Draw a right-angled triangle from these 2 points
- Determine the values of Δ**y** and Δ**x**
- Gradient = Δ**y** / Δ**x** = (50 − 28) / (3.6 − 1)
- = 22 / 2.6
- = 8.5 (1 dp)

## Rearranging equations

The **subject** of an equation is the quantity that is on its own. The **equations** that you need to recall, or choose from a sheet, often need to be **rearranged** to make a different quantity the **subject**.

- Whatever is done to one side of the equation needs to be done to the other
- **Inverse operations** are used, usually $\times$ and $\div$ **which are the opposite of each other**.

### Examples

$V = I \times R$

$V$ is the subject. Rearrange the equation to make $I$ the subject. You need to get $I$ on its own.

$I$ is multiplied by $R$, so to remove $R$ from the right-hand side, both sides of the equation need to be divided by $R$

$\dfrac{V}{R} = \dfrac{I \times R}{R}$, so $I = \dfrac{V}{R}$

$P = \dfrac{E}{t}$

$P$ is the subject. Rearrange the equation to make $t$ the subject. You need to get $t$ on its own.

$E$ is divided by $t$, so first make $E$ the subject by multiplying both sides of the equation by $t$

$P \times t = \dfrac{E \times t}{t}$ so $E = P \times t$

Now get $t$ on its own by dividing both sides by $P$

$\dfrac{P \times t}{t} = \dfrac{E}{P}$, so $t = \dfrac{E}{P}$

**Sometimes knowing the units can help you remember an equation, or you can work out a unit from an equation.**

It is much better to understand and learn how to rearrange an equation properly. However, using a **formula triangle** can help as long as you use the correct triangle. Cover the quantity you want to find and then see if you need to multiply or divide the other quantities.

| General equation type | $A = B \times C$ $\dfrac{A}{B \ \times \ C}$ | | $A = B / C$ $\dfrac{B}{A \ \times \ C}$ | |
|---|---|---|---|---|
| **Example** | potential difference = current $\times$ resistance $V = I \times R$ | | power = $\dfrac{\text{energy transferred}}{\text{time}}$ $P = E / t$ | |
| **Formula triangle** | $V = I \times R$  $I = \dfrac{V}{R}$  $R = \dfrac{V}{I}$ | | $P = \dfrac{E}{t}$  $E = P \times t$  $t = \dfrac{E}{P}$ | |

# KEY TERMS IN PRACTICAL WORK

## Experimental design

| Key term | Meaning |
|---|---|
| Evidence | Measurements or observations collected using a valid method |
| Fair test | When the dependent variable is only affected by the independent variable |
| Hypothesis | A suggested explanation for observations or facts |
| Prediction | A reasoned statement that suggests what will happen in the future |
| Valid | A valid method involves fair testing and is suitable for an investigation |
| Valid conclusion | A discussion of a valid experiment and what it shows |

## Variables

A variable is a characteristic that can be measured or observed.

| Type of variable | Meaning |
|---|---|
| Categoric | It has names or labels rather than values |
| Continuous | It has values rather than names or labels |
| Control | It affects the dependent variable, so it must be kept the same or monitored |
| Dependent | It is measured or observed each time the independent variable is changed |
| Independent | It is deliberately changed by the investigator |

## Measurements and measuring

| Key term | Meaning |
|---|---|
| Accurate | Close to the true value |
| Calibrated | A device is calibrated when its scale is checked against a known value |
| Data | Measurements or observations that have been gathered |
| Interval | The measured gap between readings |
| Precise | Very little spread about the mean value |
| Range | The values between the measured maximum and minimum values |
| Repeatable | When the same results are obtained using the same method and apparatus |
| Reproducible | Someone else gets the same results, or when different apparatus and methods are used |
| Resolution | The smallest change a measuring device can show |
| True value | The value you would get in an ideal measurement |
| Uncertainty | An interval in which the true value will be found |

## Errors

| Type of error | Meaning |
|---|---|
| Anomalous value | Anomalous results lie outside the range explained by random errors |
| Random | Unpredictably different readings – their effects are reduced by repeats |
| Systematic | Readings that differ from true values by the same amount each time |
| Zero | A type of systematic error where a device does not read 0 when it should |

# NOTES, DOODLES AND EXAM DATES

**Doodles**

**Exam dates**

Paper 1:

..............................

..............................

Paper 2:

..............................

..............................

# INDEX

## L

Laws
  Newton's 94
lenses 128
Leslie cube 124
levers 76
light 127, 130
  ultraviolet 127
  visible 122, 127, 130
limit of proportionality 72
longitudinal waves 108
loudspeakers 120, 145

## M

magnet 65, 138
  core 139
  fields 139
  poles 138
main sequence star 155
measuring wave speed 112
medical imaging 121
microphone 120, 147
microwaves 127
moment 76
momentum 99
  calculating 99, 100
  changes 101
  conservation 99
moons 155
motion in a circle 84
motor effect 143, 144, 145
moving-coil 145

## N

nebula 155
nuclear 16, 53, 54
  fission 59
  fusion 60
neutron star 156
Newton's first law 90
non-contact force 94
non-uniform motion 83

## O

orbit 157

## P

period 110
perpendicular distance 76
pistons 78
pivot 76
planets 155
poles of a magnet 138
pressure
  air 145
  at different depths 79
  atmospheric 81
  calculating 78
  in a fluid 78
principal focus 128
P-waves 121

## R

radiation
  black body 132
  infrared 124, 132
  ionising 126
  solar 133
radio waves 126
rarefaction 109
ray diagrams 117, 123, 128
reaction time 96
real images 129
red giant 156
red-shift 158
red super giant 156
reflection 118
  of waves 116
refraction 117, 118, 123
resultant force 90, 91
right hand grip rule 140
ripple tank 114
road vehicles 91
rotational effects 76

## S

satellite 155, 157
scalar quantity 83
second law of motion 91
seismic waves 121
sinking 80
Sir Isaac Newton 90
slip ring commutator 148
solar
  system 155
solar radiation 133
solenoid 140
sound waves 120
  transmission 111
spectrum
  from a star or galaxy 158
  of light 130
specular reflection 130
speed 83
spring constant 74
squares
  counting 89
stable orbit 157
star
  life-cycle 156
stars 155
stopping distance 95
stretching 72
Sun 133, 155
supernova 156
S-waves 121

## T

terminal velocity 87
thermal energy 2, 6, 8, 10, 12, 42, 98
thinking distance 95
third law of motion 94
transformers 150
transmission
  of light 131
transmission of sound waves 111
transverse waves 108

# EXAMINATION TIPS

When you practise examination questions, work out your approximate grade using the following table. This table has been produced using a rounded average of past examination series for this GCSE. Be aware that actual boundaries will vary by a few percentage points either side of those shown.

**GCSE Physics**

| Grade | 9 | 8 | 7 | 6 | 5 | 4 | 3 | 2 | 1 | U |
|---|---|---|---|---|---|---|---|---|---|---|
| F Tier (%) | | | | | 63 | 55 | 40 | 26 | 12 | 0 |
| H Tier (%) | 70 | 61 | 53 | 44 | 35 | 26 | 19 | | | |

**Combined Science: Trilogy**

| Grade | 5–5 | 5–4 | 4–4 | 4–3 | 3–3 | 3–2 | 2–2 | 2–1 | 1–1 | U |
|---|---|---|---|---|---|---|---|---|---|---|
| F Tier (%) | 59 | 54 | 50 | 44 | 37 | 31 | 25 | 19 | 13 | 0 |

| Grade | 9–9 | 9–8 | 8–8 | 8–7 | 7–7 | 7–6 | 6–6 | 6–5 | 5–5 | 5–4 | 4–4 | 4–3 | 3–3 |
|---|---|---|---|---|---|---|---|---|---|---|---|---|---|
| H Tier (%) | 66 | 62 | 58 | 53 | 49 | 44 | 40 | 35 | 31 | 26 | 22 | 19 | 14 |

1. Read questions carefully. This includes any information such as tables, diagrams and graphs.
2. Remember to cross out any work that you do not want to be marked.
3. Answer the question that is there, rather than the one you think should be there. In particular, make sure that your answer matches the command word in the question. For example, you need to recall something accurately in a describe question but not say why it happens. However, you do need to say why something happens in an explain question and should include a connecting word such as 'so', 'but', therefore', or 'because'.
4. All the examination papers will include multiple-choice questions (MCQs). Make sure you tick the correct number of boxes. When you are asked to link boxes draw straight lines. When you are asked to complete sentences using words from a word list, make sure you use words from that list.
5. Physics involves a lot of equations. Some are on the equations sheet but most need to be recalled and then used. Forgetting or failing to learn these will cost you a lot of marks. Learn the formulae well and be able to use them confidently. Also, make sure you learn a method for rearranging equations and know the correct SI units.
6. Show all the relevant working out in calculations. If you go wrong somewhere, you may still be awarded some marks if the working out is there. It is also much easier to check your answers if you can see your working out. Remember to give units when asked to do so and follow instructions about standard form or significant figures.
7. Plot the points on graphs to within half a small square. Lines of best fit can be curved or straight, but must ignore anomalous points. If the command word is sketch rather than plot, you only need to draw an approximate graph, not an accurate one.
8. Follow instructions carefully when writing or balancing nuclear equations. Check that all the numbers of particles and charges balance.
9. Remember that you may be asked to label a diagram or to complete a diagram. You may or may not be given the words to use.
10. Make sure you can recall experiments you have done or observed. About 15% of the exam is based on the required practicals.

**Good luck!**

# Revision, re-imagined
## the Clear**Revise** family expands

New titles coming soon!

These guides are everything you need to ace your exams and beam with pride. Each topic is laid out in a beautifully illustrated format that is clear, approachable and as concise and simple as possible.

They have been expertly compiled and edited by subject specialists, highly experienced examiners, industry professional and a good dollop of scientific research into what makes revision most effective. Past examination questions are essential to good preparation, improving understanding and confidence.

- Hundreds of marks worth of examination style questions
- Answers provided for all questions within the books
- Illustrated topics to improve memory and recall
- Specification references for every topic
- Examination tips and techniques
- Free Python solutions pack (CS Only)

**Absolute clarity is the aim.**

Explore the series and add to your collection at **www.clearrevise.com**

Available from all good book shops.

  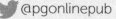 @pgonlinepub